INT...

Volume 2 Earth History

Part I Early Stages of Earth History

Introduction to Geology

Volume 2 Earth History

Part I Early Stages of Earth History

H. H. READ
F.R.S., F.R.S.E., F.G.S., D.Sc., A.R.C.S.

*Sometime Professor Emeritus of Geology,
Imperial College of Science and Technology,
University of London*

JANET WATSON
Ph.D., A.R.C.S., D.I.C.

*Professor of Geology in the University of London,
Imperial College of Science and Technology,
London*

M

First edition 1975
Reprinted 1979

Published by
THE MACMILLAN PRESS LTD
London and Basingstoke
Associated companies in Delhi Dublin
Hong Kong Johannesburg Lagos Melbourne
New York Singapore and Tokyo

ISBN 0 333 11285 7 (hard cover)
0 333 17667 7 (paper cover)

Typeset by
PREFACE LIMITED
Salisbury, Wilts
Printed in Hong Kong

Distributed in the United States by
Halsted Press, a Division of
John Wiley & Sons, Inc., New York

Library of Congress Catalog Card No. 75–501

Contents

List of Figures

List of Plates

Preface

When H. H. Read was asked by Macmillan to write an introductory textbook which would serve as the basis for a first University course in geology, he originally hoped to cover the whole subject in a single volume. It soon became clear, however, that this could not be done unless much of the factual information needed by students was jettisoned. An enlargement of the project became necessary and as I was acting as his Research Assistant he invited me to join him. We decided to think in terms of a two-volume *Introduction to Geology:* Volume 1 would deal with geological processes and with the rocks and structures produced by them, while Volume 2 would illustrate the effects of these processes by reference to the record of geological history. Volume 1 appeared in 1962 and we then began to make plans for Volume 2.

At this time the scope of historical geology was being dramatically widened by new developments in the earth sciences. It seemed clear to us that if we were to provide the background that would be needed by students in the future we must deal with earth history on a world-wide basis and must cover the full span of geological time. This broad approach made it necessary to give the book a rather unusual balance and to scale down the treatment of some important topics usually favoured by stratigraphers and palaeontologists. The plan that we adopted allowed us to deal with the geological evolution of large crustal units over long time periods. The early stages in the history of these units are covered in Part I; the later stages, from a point in time about a thousand million years ago up till the present day, are covered in Part II. Many of the themes introduced in Part I are developed in Part II, and I hope that the two Parts will be regarded as an entity for most purposes. The difficulties of producing a paperback as well as a hardcover version as a single large volume have made it preferable for Parts I and II to be published as separate but integrated books.

Soon after we had begun work on Volume 2, Professor Read suffered a severe illness which restricted his later activities. He pushed on, however, with undiminished enthusiasm and we kept in touch by means of innumerable letters and telephone conversations. About three-quarters of the manuscript had been written when he died in 1970, a few months after his eightieth birthday. Over the next year or so I completed the final chapters, revised the entire text and assembled material for the illustrations and bibliography. It falls to me, too, to thank the many people who have given us advice and information at many stages. In particular, I must mention my colleagues at Imperial College, many of them former students or colleagues of H. H. Read, who have helped to widen my horizons on many occasions. I also owe a debt of gratitude to Dr P. L. Robinson who has allowed me to use a photograph from her collection and to many geologists in Europe, Canada, Africa and Australia who have made field excursions both profitable and enjoyable. In writing a book of this scope it is impossible to keep within the limits of one's own experience and although I am, of course, entirely responsible for errors of detail, I know how greatly the book has benefited from the experience and interest of those with whom I have discussed it.

Department of Geology   JANET WATSON
Imperial College of Science
and Technology
London

Acknowledgements

The authors and publishers wish to thank the following, who have kindly supplied originals of the photographs and figures mentioned:

The Royal Air Force (Crown Copyright) Plates I and II
The Editor, *Nature, London* Figure 4.6

1

The Geological Record

I The Strands of Earth History

The rocks which make up the earth's crust are the principal documents of geological history. Every aspect of these rocks, their composition and physical properties as well as their distribution and mutual relationships, is indicative of their origin and it is the function of the geologist to establish from observation a basis for the interpretation of earth history. The crust as a whole has been changing since the earliest geological times and its present-day make-up is only intelligible when examined in relation to its past evolution. The historical approach, the distinctive trait in geological thinking, is therefore fundamental to the understanding of the earth sciences.

The geological records fall into four main groups concerned with different kinds of crustal activity. First, there are those concerned with erosion and sedimentation: the processes which co-operate to produce successions of sedimentary rocks at the earth's surface. Second, there are records of volcanic activity, the delivery to the surface or the upper parts of the crust of the molten magma from which igneous rocks are derived. Third, there are records of deformation preserved in the structures imposed on pre-existing rocks; and finally records of plutonism are supplied by the metamorphic and granitic rocks formed at depth in the crust. These diverse processes have not operated independently and the record of earth history could be likened to a fabric whose interwoven strands represent the different kinds of crustal activity.

In a previous volume, we dealt in some detail with the operation of the rock-forming geological processes. In this volume, we shall examine the interaction of these processes as recorded in the history of the crust. We shall begin by recalling some aspects which are of special relevance from the historical point of view, and from these beginnings, go on to survey the development of the earth's crust over a period of more than 3000 m.y.

II Uniformitarianism in Earth History

The pure doctrine of uniformitarianism, or actualism, states that the present is the key to the past and that geological phenomena may be interpreted in the light of observations on processes giving rise to similar phenomena at the present day. In this form the doctrine finds its greatest application in the study of erosion, deposition and effusive igneous activity at the earth's surface. The more deep-seated igneous, metamorphic and tectonic processes are less amenable to the arguments of uniformitarianism and the doctrine has to be applied to them in a somewhat different form.

The uniformitarian approach to the problems of sedimentation has made possible the interpretation of the facies of sedimentary rocks — the sum-total of their structural, petrological and faunal characters — by reference to the environments of deposition and the palaeogeography of land and sea. In the study of earth history, the facies of sedimentary rocks must clearly be of special importance. But even in regard to these rocks the uniformitarian argument has to be applied with caution. Though the style of surface processes may have remained much the same through geological history, their rates may have varied and the materials on which they worked may have changed. It has been suggested, for example, that the rate of sedimentation has increased with the passage of time, a suggestion based on comparisons of maximum known thicknesses of certain Phanerozoic systems in proportion to the duration of the corresponding geological periods.

Propositions concerning changes in the materials involved in sedimentation relate to the compositions of the atmosphere, the oceans and the continental crust itself. The atmosphere of early times is thought to have been poor in oxygen, which was added largely as a result of organic photosynthesis. This might well have influenced the processes of weathering and diagenesis, which at the present day frequently involve oxidation. The salts of the oceans — derived mainly from the weathering of surface rocks — may have had a lower concentration, which could have affected marine sedimentation and diagenesis. Still more significant factors may be involved if the crust and mantle themselves have evolved through geological time, for the bulk compositions of the igneous rocks, and therefore of the source rocks exposed to erosion, may then have suffered gradual changes. Finally, organic evolution has produced irreversible changes in conditions at the earth's surface, a circumstance that makes it difficult to interpret some ancient deposits along strictly uniformitarian lines.

When we turn to consider the history of deep-seated geological events, the problems which arise are rather different. The tectonic, magmatic, and metamorphic processes which go on at depth cannot be observed directly, except by the use of geophysical methods. Often, the best one can do is to use rocks produced by these processes in the recent past as bases for comparison with those produced in the more distant past. As might be expected, this line of approach reveals that metamorphic, migmatitic and granitic rocks of recent date can be matched in the details of their structures, textures and mineral assemblages with examples from the older crystalline terrains; such resemblances suggest that there has been no drastic alteration in the range of temperatures and pressures attained within the crust, or in the physical and chemical responses of rocks to changes of crustal environment.

The Alpine fold-belt of recent date, studied by generations of great geologists, and the somewhat older Hercynian and Caledonian belts, have been taken as patterns for fold-belts in general and many more ancient orogenic zones have been interpreted in the light of Alpine results. Such a practice may have encouraged a tendency to 'tidy-up', and force into a single pattern, the diverse phenomena connected with deep-seated orogenic activity. While it can be established with reasonable certainty that systems of mobile belts separated by stable blocks were in existence at least as far back as 2000 m.y., probably few geologists would claim that these resembled the Phanerozoic orogenic systems in every respect. The structural patterns suggest that at still earlier times (>2500 m.y.) the crust was largely mobile.

The oldest rocks of the geological record have been, at various times, the subject of non-uniformitarian proposals which attributed their production to events that were not repeated at later stages of earth history. Such proposals centre about the idea of the survival of portions of a primeval crust dating from before the start of geological processes as we know them. As the increasing numbers of radiometric age-determinations enable us to see further and further back in geological time, it becomes necessary to eliminate all but the oldest dated rocks as possible candidates for this role. As the evidence stands at present, it seems clear that only rocks older than about 3500 m.y. need be considered in this context; rocks which retain the characters acquired before this date are of restricted distribution and the possibility that some of them had a unique mode of origin does not affect the interpretation of most Precambrian areas.

There remains, however, a real possibility that the crust and mantle of the earth may have undergone a unidirectional chemical and structural evolution. So long as the attention of historical geologists was given mainly to the Phanerozoic rocks, the effects of long-term changes were difficult to detect, since the span of time involved was little more than 600 m.y. As the earlier stages of geological history become better known, new evidence may be expected to emerge which will have to be taken into account as the principle of uniformitarianism is assessed.

III Timing in Earth History

The dating of geological events relative to each other, and where possible in terms of a time-scale expressed in millions of years, is an essential step in the elucidation of earth history. The classic methods of geological timing are concerned with the dating of episodes of sedimentation, magmatism and earth-movement with regard to one another. The methods of radiometric or isotopic dating fix geological episodes with respect to the numerical time-scale. The 'clock' used in these methods is provided by the decay of radioactive elements contained in rocks or minerals. The third method of dating geological episodes depends on relating these episodes to the history of reversals of the earth's magnetic field through geological time.

Radiometric dating. Radiometric or isotopic methods of dating rocks, first tried out soon after the discovery of radioactivity, came into general use after the Second World War and are now established as standard procedures, especially for study of the first five-sixths of the geological record, where dating by means of fossils is seldom practicable.

The breakdown of radioactive isotopes goes on at constant rates. In any mineral or rock, therefore, the ratio of the parent isotope to the radiogenic isotopes should generally be proportional to age: the older the mineral, the higher is the proportion of the breakdown-product. The *half-life* of the parent element, the time taken for an initial number of atoms to be reduced by one-half, determines the span of time over which the element can be used for dating. The *decay-constant* is the fraction of a given number of atoms that decays in a stated interval of time. The principal radioactive isotopes used for dating are ^{235}U, ^{232}Th, ^{208}Pb, ^{207}Pb, ^{40}K, ^{87}Rb, ^{14}C (Table 1.1).

The interpretation of radiometric age-determinations depends to a great extent on an understanding of the history of the mineral or rock dated: to take an obvious example, detrital zircons and authigenic glauconites from a sandstone would yield dates relating to totally distinct geological events. Minerals from a metamorphosed igneous body may yield dates referring to the period of intrusion and/or to the period of metamorphism. More confusingly, they may yield *apparent ages* intermediate between two significant episodes. In igneous and metamorphic rocks formed at high temperatures, the radiometric 'clock' is set not at the moment of crystallisation but at the stage at which rocks or minerals begin to act as *closed systems*. In a rapidly cooled igneous body, the interval between consolidation and the setting of the radiometric clock may have been negligible, but in deep-seated metamorphic complexes, temperatures may have remained above the level at which the radiometric clock is set for tens or even hundreds of millions of years after the cessation of crystallisation. The apparent ages obtained from many deep-seated metamorphic rocks therefore date phases of cooling due to uplift and erosion, rather than phases of active metamorphism.

Since the methods of isotopic age-determination depend on measurement of the ratios of radioactive and radiogenic isotopes, the results may be modified by geological happening which caused one or both of the isotopes to migrate through the rock-system. Argon and helium are, not unnaturally, prone to wander and techniques involving these gases may therefore yield anomalous results. More generally, the lattice-structure and grain-size of the host mineral and the fabric of the host rock may influence the mobility of the isotopes. Geological events which promote migration include tectonic disturbance, rise of temperature and hydrothermal activity. Rocks in the vicinity of faults often yield anomalous apparent ages. A 'thermal event' insufficient to bring about metamorphic recrystallisation may drive off radiogenic isotopes. A pattern of apparent ages obtained by the use of different techniques or by the dating of individual minerals and of whole-rock samples from a single geological unit may be spread over a considerable time-range as a result of such disturbances.

Timing of reversals of the earth's magnetic field. Reversals of the earth's magnetic field have taken place repeatedly during at least the latter parts of geological history. The spacing of magnetic reversals can be used, like other

Table 1.1. RADIOMETRIC METHODS OF AGE-DETERMINATION

1. Uranium and thorium methods

$$^{238}U \rightarrow {}^{206}Pb + 8\ {}^{4}He$$
$$^{235}U \rightarrow {}^{207}Pb + 7\ {}^{4}He$$
$$^{232}Th \rightarrow {}^{208}Pb + 6\ {}^{4}He$$

Application: dating of uranium minerals and also of minerals such as zircon and monazite containing small amounts of U or Th. Zircon ages often refer to the date of formation of a rock, since zircon is resistant to change. The *concordia method*, which depends on determining isotopic ratios in a group of related samples, may give a reliable date of formation.

2. Common lead methods

   Application: to galenas, dates time of separation of lead from sources containing U and Th and may therefore give dates much older than time of emplacement of galena; to whole rocks containing lead as a trace element.

3. Potassium—argon methods (K—Ar)

$$^{40}K \rightarrow {}^{40}A$$

Application: a versatile method applicable to many igneous and metamorphic rocks and to sediments carrying glauconite, illite etc. K-feldspars and micas provide suitable material and minerals and whole-rocks low in K can also often be dated by this method. Low apparent ages may result from argon loss due to late thermal events: anomalously high ages from the accumulation of excess argon in certain minerals. An 'age-spectrum' which may identify several events is obtainable by special techniques.

4. Rubidium—strontium methods (Rb—Sr)

$$^{87}Rb \rightarrow {}^{87}Sr$$

Application: a versatile method, since Rb is a widely distributed trace element in alkali-feldspars and micas. The *isochron method*, which depends on determination of the ratios ^{87}Rb: ^{86}Sr and ^{87}Sr: ^{86}Sr in a number of genetically related rocks and minerals may indicate date of formation of igneous rocks with complex histories.

5. Radiocarbon methods

$$^{14}C \rightarrow {}^{14}N$$

A method used in the dating of young organic material, especially valuable in archaeology. Since the half-life of radiocarbon is short (5570 years) the method can be applied only to materials less than 100 000 years in age.

recurrent happenings, as *time-markers* to which episodes of geological history can be referred. This framework of magnetic reversals has proved to be of critical importance in establishing the history of the crust in oceanic basins.

With a few exceptions, igneous rocks containing ferromagnetic materials and sediments containing ferromagnetic detrital particles or cement become magnetized at the time of formation in the direction of the earth's field. Systematic investigations have shown that rocks formed during certain geological time-intervals have consistently reversed magnetization, while those of the inter-vening periods are normally magnetized. Roughly half the analysed Phanerozoic rocks show normal magnetization and half are reversed. Although a very few instances are known in which reversals appear to be controlled by the geochemical and mineralogical properties of the rocks concerned, the reality of the periodic reversals of the earth's field appears to be unquestionable (see e.g. Bullard, 1968).

The recognition and dating of magnetic reversals have not yet been completed even for the Phanerozoic eon. The most reliable results, so far, have been obtained from lavas less than four million years in age which can in favourable circumstances be dated to within about 100 000 years by K–Ar methods. Virtually all Recent and Pleistocene rocks are normally magnetized and define the latest period of normal magnetization. Going back in time, four major periods of alternately normal and reversed magnetization, each interrupted by several short reversals, have been identified over 4 m.y. (Fig. 1.2). During the

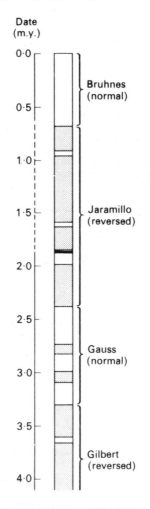

Date
(m.y.)

0·0

Bruhnes
(normal)

0·5

1·0

1·5   Jaramillo
       (reversed)

2·0

2·5

Gauss
(normal)

3·0

3·5

Gilbert
(reversed)

4·0

Fig. 1.1. The sequence of magnetic reversals derived from volcanic rocks set against a time-scale established by K–Ar dating of these rocks (based on Doell *et al.*, 1966) (see page 000)

whole of the Tertiary era, some seventy major periods are defined by magnetic reversals. The spacing of reversals in both Tertiary and pre-Tertiary periods appears to have fluctuated rather widely: for example, rocks formed during the 55 m.y. of the Permian period have proved to be almost entirely reversed.

IV Divisions of the Stratigraphical Record

The Cryptozoic and Phanerozoic eons. The geological history of the first five-sixths of geological time is recorded in rocks which contain almost no fossils and which are frequently deformed or metamorphosed, or both. Only for the final 600 m.y. does an adequate fossil record provide the basis for stratigraphical correlation. Although this contrast is not of fundamental importance so far as the history of the crust itself is concerned, it has profoundly affected geologists' understanding of that history. It is often convenient, therefore, to divide the record at the point where fossils first appear widely, that is, immediately below the Lower Cambrian deposits characterised by the *Olenellus* or *Archaeocyathus* faunas. We have adopted for the two 'eons' thus distinguished, the names proposed by Chadwick:

2 *Phanerozoic eon* ('obvious life') = Cambrian and later
1 *Cryptozoic eon* ('hidden life') = pre-Cambrian

These terms are used when it is necessary to express the antithesis between the two contrasted records. For ordinary purposes, we shall employ the more usual term *Precambrian* for the rocks formed during the Cryptozoic eon.

The stratigraphical column. The sequence of Phanerozoic sedimentary rocks, arranged in a continuous column from oldest at the bottom to youngest at the top, provides, by long tradition, the favoured material for the studies of the stratigrapher and the palaeontologist. The sequence of Precambrian sedimentary rocks has not yet been established in its entirety and the construction of a Precambrian stratigraphical column remains a task for the future (see p. 138).

The ideal stratigraphical column provides a continuous history of sedimentation and, as with human history, no world-wide divisions can be made in it. Not all the divisions adopted in practice are based on comparable criteria, some being made by reference to regions which were studied by early geologists and others being more purely palaeontological. Side by side with these divisions is the time-scale in terms of millions of years, which is gradually being linked to the classic stratigraphical column by radiometric dating of identifiable horizons.

In the early days of systematic geology, most major boundaries were drawn at unconformities that were conspicuous in western Europe; but no unconformity and no orogenic upheaval responsible for it has turned out to be of world-wide significance. Again, epeirogenic movements have been invoked and a transgression and regression proposed as a convenient time-unit – but here also the application is not of world-wide validity. Faunal and floral data seem, on the whole, to provide the best basis for the definition of even the major divisions.

The fundamental differences in the fossil-assemblages of the early, middle and late Phanerozoic rocks gave rise to the old-established threefold division into Palaeozoic, Mesozoic (or Secondary) and Cainozoic (Cenozoic or Tertiary) eras – the eras of old, middle and recent life. The independence of the so-called

Quaternary era can only be defined by reference to the unique character of one member of its fauna, man. The principal divisions of time within each era constitute the geological *periods* during which the rocks of the corresponding *systems* were laid down. Smaller time-divisions correspond to *series* and *stages* of the stratigraphical column as shown in Table 1.2. The placing of boundaries between such divisions must in the end be arbitrary: the boundaries may finally be defined by reference to fixed points in internationally accepted *type-sections* of each portion of the column. Where the time-significance of lithological divisions is unknown, it is customary to employ the *rock-stratigraphic* terminology summarised in the lower part of Table 1.2.

Table 1.2. STRATIGRAPHICAL TERMINOLOGY

1 *Time-stratigraphic terms* in use for the purposes of equating time-periods with rock-sequences:

Rock-unit	Equivalent time-unit
	eon
	era
system	period
series	epoch
stage	age

2 *Rock-stratigraphical terms* in use for the purposes of local or regional descriptions of rock sequences. A rock-stratigraphic (or lithostratigraphic) unit is recognised and delimited on the basis of lithological characteristics: the boundaries of such units may be diachronous. The principal terms, in decreasing order, are:

Supergroup:	two or more related groups
Group:	two or more related formations
Formation:	the mapping unit, the basic practical unit
Member	
Bed	

(for detailed discussion see the 'Code of Stratigraphic Nomenclature', *Bull. Am. Ass. Petrol. Geol.*, (1961) **45**, 645–65. Report of Stratigraphical Code Sub-Committee (*Proc. geol. Soc. Lond*, 1967, No. 1638)

Direct determinations of the ages of sedimentary rocks have been derived from the dating of authigenic glauconite and of illite and sylvine by K—Ar methods and from whole-rock K—Ar dating of shales. Trustworthy determinations may also be obtained from volcanic intercalations, and more ambiguous evidence from the dating of igneous rocks intrusive into the succession under investigation.

The rates of sedimentation. The dating of fixed points in the Phanerozoic succession has provided new means of estimating the rates at which the processes of deposition operate. These rates vary enormously according to the sedimentary facies and the agent of deposition. The abyssal red clay is believed to accumulate at a rate of only a millimetre or so in a thousand years, while several metres of sediment may be laid down over a few days by major turbidity currents. Spectacular agents of deposition such as turbidity currents and rivers in flood are

Table 1.3. RATE OF DEPOSITION (Based on Sutton, 1969)

Locality	Period	Subsidence (km)	Rate (km/m.y. or mm/yr)
Appalachian Mountains, Newark Sandstone	U. Triassic 25 m.y.	6	0.24
East Greenland, Old Red Sandstone	M. Devonian 12 m.y.	4	0.33
Fossa Magna Japan	Miocene 19 m.y.	15	0.78
Lake Baikal U.S.S.R.	Tertiary—Recent 25 m.y.	7	0.28
Gulf of Mexico	Pleistocene—Recent 2 m.y.	1.5	0.75
	L. Miocene—Recent 25 m.y.	7	0.28
	U. Jurassic—Recent 150 m.y.	14	0.09

often intermittent in their action, and rates of accumulation must therefore be averaged out over a period of time (Table 1.3). Maximum rates obtained for a number of Palaeozoic successions are in the region of 3—4 cm per 100 years. It should be emphasised that these rates refer to successions of quite exceptional thickness; the greatest thicknesses of sedimentary rocks laid down in mobile belts during a single geological period seldom rise much above 8 km.

V The Integration of Crustal History

The development, evolution and eventual stabilisation of successive systems of mobile belts in the crust impose a kind of pattern on the geological record which we shall use as a basis for the systematic descriptions which comprise the main part of this book. The fact that mobile belts have developed, decayed, and been replaced by others over periods to be measured in hundreds of millions of years becomes apparent when the geological history of any continental mass is surveyed. The Phanerozoic history of western Europe, for example, records the development of three principal systems – one of early Palaeozoic date in north-west Europe, one of late Palaeozoic date in central and southern Europe and a third of Mesozoic and Tertiary date in the Mediterranean regions (Fig. 2.1). Elements of a common pattern involving sedimentation and igneous and plutonic processes as well as earth-movement can be made out in all three and are repeated, though with characteristic variations, in the history of many orogenic mobile belts, providing the natural basis for a first division of the geological record.

This manner of approaching the study of earth history appears to be valid, in a very general way, for the greater part of geological time and has the further advantage that it takes into account the records of deep-seated processes as well as those of the surface events which provide the traditional stuff of stratigraphy. In adopting it we follow lines pioneered by J. J. Sederholm (1863—1934) and by Arthur Holmes (1890—1965). It was Sederholm who first distinguished

successive Cryptozoic fold-belts. His work among the migmatitic and metamorphic Precambrian rocks of Finland was used by Holmes as a starting point
for his own studies of the Cryptozoic fold-belts of east and central Africa.
Holmes concluded that each region of Precambrian rocks has 'an ingrained
pattern of successive orogenic belts and each continent appears to be an
integration of many such belts.' Successive belts could be dated by structural
criteria combined with radiometric determinations of rock-ages. Holmes' first
application of these principles was at variance with current classifications of
Precambrian rocks in Africa, but although subsequent radiometric age-determinations have modified the details of his proposals, his methods have been
completely vindicated.

The combined methods of establishing regional equivalence provided by
geological and geochronological methods enable us to take the mobile belt as a

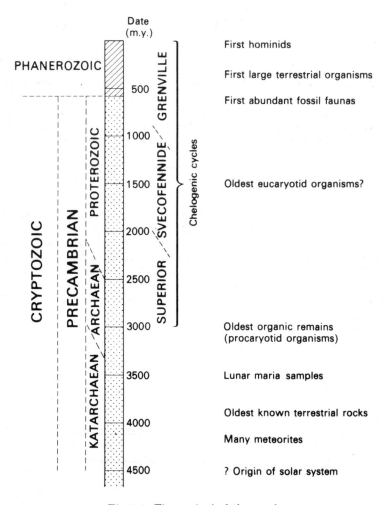

Fig. 1.2. The geological time-scale

principal unit of geological structure and the evolution and stabilisation of successive belts as the means of marking-off some broad time-divisions in geological history. These units provide the framework into which details of every kind can ultimately be fitted.

The life-span of the mobile belt. The time which elapsed between the deposition of the earliest sediments related to a mobile belt and the erosion of the stabilised belt is the period which Sutton has called its *life-span.* The duration of individual phases can be estimated in relation to the whole life-span by dating the earliest and latest geosynclinal sediment, the earliest and latest episodes of folding and metamorphism and so on. Figure 1.3, summarising information of this kind is based on estimates of Sutton. The overall life-span of the belts mentioned varies from 200 m.y. to over 600 m.y. Episodes of orogenic folding and metamorphism may recur over periods ranging from 30 to 60 per cent of the life-span. As will be seen later, some Precambrian belts appear to have remained mobile over even longer life-spans. These profound differences have to be taken into account in any discussion of the ultimate mechanism controlling crustal processes.

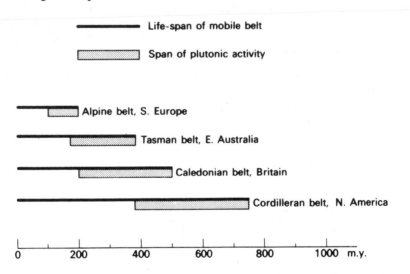

Fig. 1.3. Variations in life-span of mobile belts (based in part on Sutton, 1963)

VI Grouping the Geological Cycles

Variations in plutonism. In 1960 Gastil published a histogram showing the time-distribution of mineral dates then available from all over the world and concluded that the pattern of peaks revealed by it suggested lengthy fluctuations in the earth's orogenic history. This conclusion, based on a small sample, has been endorsed in a general way by several later authors who have been able to use a larger number of age-determinations. Gastil's original diagram revealed seven peaks separated by intervals poor in, or lacking in, determinations. A diagram published more recently by Dearnley (Fig. 1.4) shows a simplification which suggests that some of the gaps were due to lack of data. Nevertheless, several low points persist, notably about the times 2100 ± 200 m.y., 1200 ± 100 m.y. and 800 ± 100 m.y.

The pattern revealed by these diagrams from world-wide sources is naturally a composite one and the fact that it exhibits marked fluctuations in itself suggests that long-term variations in plutonic activity have been broadly contemporaneous over the whole world. The same point is made more clearly when data from individual continents are plotted separately (Fig. 4.3). The principal low points on each histogram fall at roughly the same positions on the time-scale, though the intervening peaks do not always coincide exactly.

It is important to recall the basis of these diagrams. The number of radiometric determinations relating to any period or region is subject to accidental variations resulting from such factors as the accessibility of the regions to be sampled. Early episodes in long-lived mobile belts are under-represented because their effects have been obliterated by subsequent recrystallisation. In spite of these defects, we consider that the diagrams reflect real fluctuations in plutonic activity. The low points and the slopes linking these to the preceding peaks, as Sutton has pointed out, mark periods when plutonic activity was drastically reduced and when regional uplift led to general cooling. They may therefore be the most significant features of the diagrams. The peaks indicate in a general way the periods of more widespread activity, though their

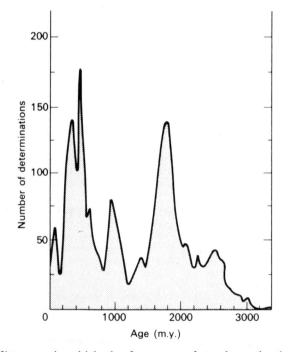

Fig. 1.4. Histogram in which the frequency of age-determinations for granitic and metamorphic rocks is plotted against the time-scale (3400 determinations, after Dearnley, 1966)

highest points may mark the termination of the last, rather than the most extensive, phase of plutonism in a particular belt. The periods 2800–2600 m.y., 1900–1600 m.y., 1150–900 m.y. and 500–0 m.y. are indicated as times of considerable plutonic activity.

From the foregoing discussion we can turn to the problem of subdividing the length of earth history on the basis of a rhythm of mobility which may be termed the orogenic or geological cycle. It is immediately evident that successive cycles overlapped in time. Within the Phanerozoic eon, for example, the initial phases of deposition in the Hercynian mobile belt began well before the final phases of molasse deposition and vulcanicity had come to an end in the Caledonian belt. By reference to the temporal fluctuations of plutonic activity here discussed and the evidence of periodic shifts in the sites of activity referred to later, we adopt a fourfold grouping of cycles as shown in Table 1.4. The major

Table 1.4. GROUPING THE GEOLOGICAL CYCLES

	Completed before	Age terms	Chelogenic cycle	Tectonic regime
D	not completed	Phanerozoic late-Proterozoic	Grenville	Grenville
C	1100 m.y.	Proterozoic	Svecofennide	Hudsonian
B	2100 m.y.	Archaean	Shamvaian	Superior
A (oldest)	3000 m.y.	Katarchaean		

time-periods distinguished in the table are marked off roughly by the low points on the histograms of radiometric dates. Each period covers the overlapping life-spans of a number of systems of mobile belts rather than the life-span of a single belt and is therefore capable of further subdivision. Several preliminary comments on this table must be made. In the first place, events dating back to the earliest recorded phases of earth history (say, prior to 3100 m.y.) are too sparsely represented by datable rock-material to figure in the histograms; these phases are represented in the table by reference to other considerations. In the second place, it makes no distinction between successive Phanerozoic geological cycles, which are all lumped together in the third group. This is a natural effect of the system adopted, which lays emphasis on an extremely slow rhythm in crustal activity. It can be compensated for by subdivision of the major groups where appropriate. Finally, the table introduces several terms which have not hitherto been discussed. The names *Katarchaean, Archaean* and *Proterozoic* are regarded as time-terms, although their limits have not yet been precisely fixed (see Fig. 1.2). They will be used throughout this book and their validity discussed later (p. 178). The term *chelogenic cycle* was proposed by Sutton in 1963, for the long time-periods of the fundamental rhythm and the term *tectonic regime* (Dearnley, 1966) expresses a similar concept. This term is derived from considerations which have still to be discussed, namely the distribution of groups of mobile-belt systems in space.

Orogeny in space. The distinction between mobile and stable regions that has characterised the crust for much of its history persists, with some shifts and variations, through the length of a geological cycle. The ending of a cycle brings stability to tracts which were formerly mobile and the initiation of a new cycle is marked by the onset of mobility in tracts which were formerly stable. Fluctuations of crustal mobility are therefore associated with changes in the distribution of areas of mobility, which provide a basis for the recognition of some major crustal units.

A *tectonic province* constitutes the field of operation of a single geological cycle (e.g. Quennell and Halderman, 1960) or of a group of cycles closely linked in time. As defined by Quennell and Halderman (p. 173) the province 'includes all that region which is affected by the tectonics, erosion, sedimentation, plutonism and vulcanism of the cycle' but we will restrict the term as do Canadian geologists, to terrains which are physically continuous. Regions affected by the same cycle, which are separated by younger mobile belts or by ocean basins, are therefore to be assigned to separate provinces. A young tectonic province usually represents almost the whole extent of the corresponding mobile belt within any continent, whereas older provinces usually represent only those remants of the original belts which have escaped burial or remobilisation.

According to G. W. Tyrrell, a *shield* is a block, often of continental dimensions, which has not undergone tectonic disturbance, apart from gentle vertical movements, since early geological time. Combined with this is the requirement that it had at that earlier time suffered deformation, metamorphism and granitisation which converted it into a strong resistant block (McConnell, 1948). Most shields resolve themselves, when mapped in detail, into a number of Precambrian tectonic provinces. Putting together these various points, we arrive at the conclusion that a shield represents the stabilised portions of Precambrian mobile belts welded into a stable block and exposed as a result of persistent uplift through Phanerozoic times. The more general term *craton* includes stable platforms which carry a blanket (cratonic cover) of younger deposits, as well as shields devoid of an undisturbed cover.

For the purposes of dealing with the early parts of geological history, the great shields (*boucliers* is the French equivalent employed by Suess) provide the natural divisions on which to base a regional description. We group the principal shield areas in two major clusters:

LAURASIA: Canada, Greenland, Baltic, Siberia.
GONDWANALAND: Africa, Peninsular India, South America,
                           Australia, Antarctica.

This grouping immediately introduces two further regional units whose importance is now well-established in geological thinking. *Laurasia* denotes, broadly speaking, the continental masses of the Northern Hemisphere: its name is derived from the St Lawrence River of Canada and from Eurasia. *Gondwanaland* denotes the continental masses of South America, Africa, Arabia, India, Australia and Antarctica which lie south of, or close to, the equator; its name, given by Suess, is derived from the primitive Gonds of central India. The components of Laurasia and Gondwanaland are separated by the meridional Alpine-Himalayan fold belt of Mesozoic and Tertiary age and it has long been recognised that those of each group are united by certain common features in their Phanerozoic histories. In mid-Phanerozoic times the components of

Laurasia and Gondwanaland were clustered together in two 'supercontinents' perhaps loosely connected, which were subsequently fragmented and dispersed as a result of continental drift. In dealing with the Precambrian and early Phanerozoic history of the supercontinents, allowance will be made for the effects of late Phanerozoic disruption and drift.

VII The Scheme Adopted

The succeeding parts of this book provide an outline of earth history arranged on a regional basis. Part I deals with the events related to those geological cycles which were completed in Precambrian times. It is concerned mainly with the rocks of the shields which have remained stable since the end of the Precambrian, and it ends with a general review of the evidence relating to the early evolution of the earth's crust.

In Part II, we consider the evolution of mobile belts and stable regions formed over the period from 900 m.y. ago to the present day, that is, during late Precambrian and Phanerozoic times. Most attention is given in the early chapters to three well-known belts, the Caledonides and Hercynides of Europe, and the Appalachians of North America, while other belts formed over the same time-span are dealt with more briefly. In the later chapters, accounts of the western mobile belt and stable cratons of the Americas and of the Alpine-Himalayan mobile belt and of the cratons of Eurasia and Gondwanaland bring the history of the continental masses up to date. The development of ocean basins over the Mesozoic and Tertiary eras — the first periods for which direct geological records of the history of oceanic regions are available — is dealt with in two chapters covering the new-formed Atlantic and Indian Oceans and the old-established Pacific Ocean.

We should perhaps make clear that the abundant geological information relating to the Phanerozoic periods has been deliberately scaled down in an attempt to bring our treatment of the later stages of earth history into line with that adopted for the earlier stages. At the risk of considerable oversimplification, we hope in this way to allow the emphasis to fall on those aspects of earth history which can be traced through the full span of geological time.

2

European Shield-Areas

I Preliminary: the Structural make-up of Europe

We will preface the account of the shield-areas of each continent with a short summary of the make-up of the continent as a whole, illustrated by a diagrammatic map for reference in later parts of the book. Accordingly, Fig. 2.1 shows the principal structural units of Europe in which the main events in the history of the continent are recorded.

The first unit is the ancient *Baltic shield*, made up of rocks whose history was completed about 1000 m.y. ago. This shield lies at the northern border of the continent and passes south-eastward beneath a cover of little disturbed Phanerozoic rocks which make the *Russian platform* of eastern Europe and western Asia. Precambrian rocks appear again further south in the *Ukrainian massif* and in scattered outcrops in the valleys of the Dnieper and the Bug, north of the Black Sea. These shield and platform areas together constitute a huge *European craton*, which has remained more or less stable throughout Phanerozoic time. A fragment of another ancient stable region, preserved in north-west Scotland, is a part of the *North Atlantic shield* which has been disrupted by continental drift: a larger portion now forms Central and West Greenland (see Chapter 3).

The *Caledonian system of orogenic belts*, developed over the period 900–400 m.y., constitutes the third component of the continent. Its principal outcrop extends for 2000 km, from western Ireland to the North Cape, and makes the Caledonides of Britain and Scandinavia. Other outcrops, probably detached from this belt by continental drift, extend along the north and east coasts of Greenland and through Spitsbergen. Traces of an early Caledonian belt have been identified in central Europe, where they have been largely reworked during later orogenic activity.

The main *Hercynian or Variscan system of orogenic belts*, built up over the period 400–280 m.y., occupies a very broad zone extending roughly east and west through the southern and central parts of Europe. A little disturbed post-Hercynian cover-succession hides much of the northern part of the orogenic zone; the southern part is extensively modified by Alpine orogenic activity. The

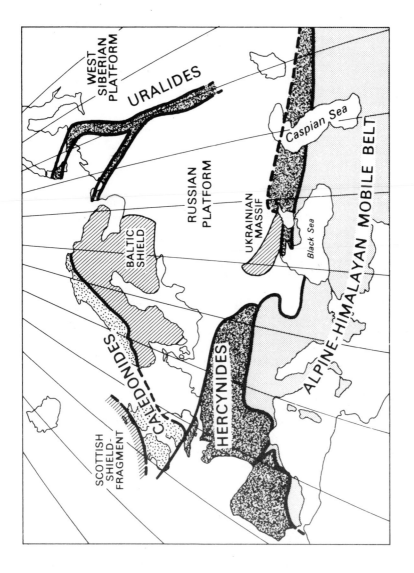

Fig. 2.1. The main tectonic units of Europe and western Asia

orogenic belt of *the Urals*, running roughly north and south and also stabilised towards the end of the Palaeozoic, separates the Russian platform from the Siberian platform and can conveniently be taken as the eastern limit of the area under consideration.

The *Alpine system of orogenic belts*, developed over the period 300–0 m.y., constitutes the last great structural unit of Europe. This system, superimposed on the Hercynian belt, forms an intricate pattern of arcuate belts crossing southern Europe, the Mediterranean region, and north Africa, in a general west-to-east direction.

II The Baltic shield

The Baltic shield, occupying most of Sweden and Finland, southern Norway and the region of Karelia in north-western U.S.S.R., has an area of at least 500 000 sq km. It is limited on the south simply by the edge of the cratonic cover, and evidence from boreholes and from geophysical surveys shows that the structures of the shield are continued in the basement of the Russian platform (Fig. 2.3). To the north-west, on the other hand, the shield is abruptly truncated by the front of the Caledonian mobile belt within which Precambrian rocks have been extensively reworked.

The Baltic shield is perhaps better known than any other Precambrian area of similar size and provided the testing ground in which new methods for the investigation of crystalline complexes were tried out in the early decades of the twentieth century. The work of J. J. Sederholm in Finland led him to interpretations of Precambrian geological history in terms of the effects of successive orogenic events and paved the way for the Swiss geologist Wegmann to apply his experience of the Alps. Sederholm's concept of the process of migmatisation was followed up by Eskola, Wegmann and Backlund in a series of studies which demonstrated the feasibility of treating migmatites as documents of geological history. Their methods have become standard practice.

The Baltic shield (Fig. 2.2 and Table 2.1) is made up of three major tectonic provinces. The greater part of the shield is occupied by the *Svecofennide system of mobile belts* which was stabilised in early Proterozoic times. To the north and east of these belts lies a province of older rocks which remained stable throughout the Svecofennide cycle. To the west lies a province representing part of a later Proterozoic mobile belt much of which is now incorporated in the Caledonian belt. Some of the common terms used in the literature are noted for reference in Table 2.1. The principal chronological groupings are:

(a)   *The Archaean province*
    I   Katarchaean remnants 3500–3000 m.y.
    II   The Saamide belt 2870–2150 m.y.
    III   The Belomoride belt 2100–1950 m.y.

(b)   *The Svecofennide province*
    IV*a*   The Svecofennides
    IV*b*   The Karelides   2000–1760 m.y.

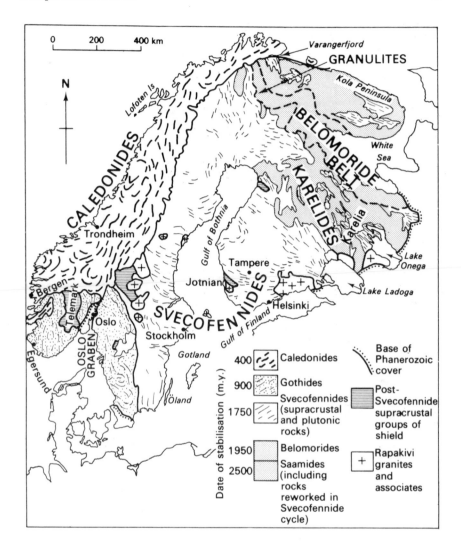

Fig. 2.2. The Baltic shield (based on compilations by Simonen

(c) *The Riphean province*
V   1200–900 m.y.

The termination of orogenic mobility in the Svecofennide system marked a
turning point in the evolution of the shield. From this time onward the greater
part of the shield remained stable and its history is recorded by fragments of a
cratonic cover-succession and by anorogenic igneous suites.

Table 2.1. PRECAMBRIAN UNITS OF THE BALTIC SHIELD

	Fold-complexes, mobile belts etc.	Major stratigraphical units, supracrustal groups	Igneous suites
LATE PROTEROZOIC	Caledonides (initial stages) Riphean Sveconorwegian ? Gothides	Sparagmites (Infracambrian) Telemark Dalslandian Jotnian sandstone	Hyperites of Riphean belt Post-Jotnian diabases of craton
EARLY PROTEROZOIC	Karelides Svecofennides	Svecofennian Karelian Lapponian Svionian	Rapakivi granites, Dala porphyries Svecofennide } orogenic Karelide } suites
ARCHAEAN	Belomorides Saamides		
KATARCHAEAN	Katarchaean remnants		

III The Ukrainian Massif

The Ukrainian massif (Fig. 2.3), with an area of 200 000 km^2, reveals fragments of Archaean and Proterozoic provinces framed by the outcrop of a cratonic cover which, though composed mostly of Phanerozoic rocks, locally includes late Proterozoic members. A sequence of 'fold-complexes' has been recognised in the massif itself by Soviet geologists; Semenenko (1968) gives the following groupings:

(a) Katarchaean 3600–2700 m.y.
(b) Lower Archaean (Dnieprovian) 2700–2300 m.y.
(c) Upper Archaean (Bug–Podolian) 2300–1900 m.y.
(d) Lower Proterozoic (Krivoi Rog) 1900–1700 m.y.
(e) Upper Proterozoic (Ovruch and Volnian) 1700–1150 m.y.

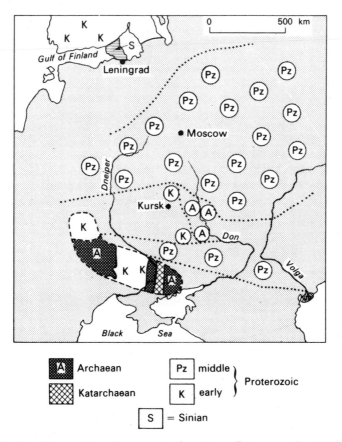

Fig. 2.3 The Precambrian basement of eastern Europe and western Asia. Symbols enclosed in circles refer to the basement beneath the Phanerozoic cover of the Russian platform

The effects of successive cycles appear to be superimposed on one another in a manner which makes it difficult to distinguish clearly defined tectonic provinces. The Archaean and Katarchaean units are preserved mainly as relict masses enclosed in a belt showing the effects of Krivoi Rog folding, while the later Proterozoic divisions occupy the north-western part of the massif.

IV The Archaean Provinces

Complexes that have remained unmodified since the beginning of the Sveco-fennide cycle in earliest Proterozoic times are preserved in the main Archaean province of the Baltic shield, which constituted the eastern foreland of the Karelide belt, and in a number of smaller massifs enclosed within the mobile belts of the Svecofennide-Krivoi Rog system. Rocks of similar ages occurring in the Proterozoic provinces are usually strongly modified.

1 Katarchaean remnants

The oldest rocks so far identified in the Baltic shield are granites, gneisses, pegmatites and micaschists which crop out over an area of about $600 \, km^2$ within the terrain occupied by early Archaean complexes. These granites and gneisses have yielded isotopic ages of 3500–3000 m.y. In the Ukrainian massif, a greater range of rock-types yielding ages of 3600–3000 m.y. has been identified. These include not only granites, migmatites and gneisses but also low-grade metamorphosed basic and ultrabasic rocks and metasediments, among which are ferruginous cherts similar to those which characterise the Krivoi Rog division (p. 34). None of the Katarchaean remnants appears to occupy an area of any great size. They are regarded as being enveloped in regenerated gneisses and represent kernels that escaped reworking during later phases of Archaean plutonism.

2 The Saamides and the Dneiprovian blocks

Next in order of age come the more extensive assemblages, which yield isotopic ages in the range 2700–2300 m.y. These assemblages, represented by rather monotonous granites and granitic gneisses with remnants of highly-altered metasediments and metavolcanics, crop out east of the Karelide belt in Karelia and provide the basement on which the early Proterozoic Karelian succession accumulated. Rather similar rocks envelop the Katarchaean remnants of the Kola peninsula, which again formed part of the foreland of the Karelides. In the Ukrainian massif, granites, gneisses and highly altered supracrustals found in association with the kernels of Katarchaean age mentioned above retain some elements of an early structural pattern, which is distinct from that of the enveloping Krivoi Rog belt.

3 The Belomorides

The Belomoride belt, extending northwestward from the White Sea into Lapland, represents the oldest portion of the Baltic shield that can be regarded as a geological entity (Fig. 2.2). It is preserved largely as a median massif, flanked on either side by branches of the younger Karelide system; its original margins are therefore obscure but the occurrence of still older complexes in Karelia and the Kola peninsula suggests that, at least in its final stages, it constituted a relatively narrow mobile tract flanked by more stable areas. The Belomoride belt and the Bug-Podolian complexes of the Ukrainian massif are thought to have been stabilised over the period 2100–1950 m.y. and to have formed part of the basement on which Karelian and Krivoi Rog supracrustals accumulated. We may note in passing that this timing is somewhat unusual – in most Archaean provinces, stabilisation was completed by 2200 m.y. or even earlier. Soviet geochronologists have recently reported K–Ar dates of as high as 2800 m.y. from within the belt, suggesting that it had a long history of mobility.

The Belomoride tract contains considerable areas of highly-metamorphosed and migmatised rocks derived from supracrustal series which include quartzites, amphibolites and pelitic gneisses and schists carrying kyanite and garnet, or sillimanite and cordierite. In Lapland, supracrustals and early intrusions form a distinctive assemblage of granulite facies, the *Lapland granulites*, characterised by the presence of garnet, cordierite, sillimanite and hypersthene and by the occurrence of exceptionally strong planar fabrics. The tectonic trends in the granulite massif are arcuate, with low inward dips to north and east; those of the remainder of the Belomoride tract are generally north-west—south-east, with predominant north-easterly dips complicated by later warping.

V The Svecofennide Province

The large early Proterozoic tectonic province of the Baltic shield, to which we apply the general name *Svecofennide province*, occupies most of south-west Finland and Sweden (Fig. 2.2). Its eastern and northern borders mark the tectonic front separating the Svecofennide system of mobile belts from the older complexes that represent the foreland. The life-span of the Svecofennide mobile tracts appears to have occupied the time between the ending of activity in the Belomorides (c. 2100–1950 m.y.) and the stabilisation of the Svecofennide province itself at about 1700 m.y. The majority of isotopic dates both for metamorphic rocks and for orogenic igneous rocks in the province fall in a rather narrow range, c. 1950–1500 m.y.

The Svecofennide province is made up of two rather different tectonic entities – the *Svecofennide belt proper*, which crosses southern Finland and Sweden from east to west, and the *Karelide belt*, which runs north-westward from Karelia to the head of the Gulf of Bothnia. The relationships of these two belts have been much debated. Sederholm regarded the Svecofennide belt as an older structure truncated by a younger Karelide belt. Isotopic dates for late-orogenic granites and metamorphic minerals from both belts, however, fall within the same time-range (1850–1750 m.y.). Re-examination of the critical region near the border of Finland and U.S.S.R., in which the two belts meet, appears to have convinced Simonen and his colleagues of the Finnish Geological Survey that the east-west tectonic trends of the Svecofennides turn sharply into parallelism with the Karelide trends and that the Karelian cover-succession is not uncomformable on rocks affected by Svecofennide orogenic activity. We shall follow these authorities in regarding the Karelides and Svecofennides as components of a single orogenic system. It would be difficult, however, to exclude the possibility that the Svecofennide belt may have had a more complex history than that of the Karelides.

VI The Svecofennides proper

The dominant rocks of this belt are metasediments and metavolcanics making up the *Svecofennian succession*, together with migmatites developed from these rocks and igneous bodies intruded into them. A pre-Svecofennian basement has not been identified and in this respect the Svecofennides contrast with the Karelides.

1 The Tampere succession

The Svecofennian supracrustal rocks occupy a number of tracts winding between large bodies of migmatitic or igneous plutonic rocks whose outlines often determine the local trend. Their original characters are well preserved in the *Tampere schist belt* in Finland where the grade of metamorphism is low. This belt is bounded on the north by a group of large plutonic masses, while to the south and south-west its rocks become more highly metamorphosed and migmatised, so that their original characters are less easy to establish. The standard succession is that of the Tampere region itself (Table 2.2) but there are

Table 2.2. THE SVECOFENNIAN SUCCESION OF TAMPERE
(based on Simonen, 1960)

		Thickness (metres)
Upper Svecofennian	dominantly pelitic (not represented in the Tampere district itself)	
Middle Svecofennian	basic volcanics	1000
	conglomerates, greywacke-slates and arkoses	7800
	basic and intermediate volcanics	800–1500
Lower Svecofennian	Quartzofeldspathic rocks (arkoses, greywackes, pyroclastics)	1500–2200
(no base known)	greywacke-slates	3000

indications of lateral variations as this succession extends southward to the Gulf of Finland and westward into Sweden.

It will be seen that the assemblage is of a highly unstable facies. The sedimentary members are largely turbidites of greywacke facies. Graded bedding is a characteristic structure, the individual beds varying from sandy units a metre or so in thickness, to layers only a few centimetres thick, which pass up from fine silt at the base to thoroughly pelitic material at the top. Other sedimentary structures in the greywackes include sole-markings, 'candle-flaming' of the pelitic layers into peaks drawn up into the base of the overlying bed and the occurrence of shale fragments in graded beds.

Minor members of the sedimentary groups are black shales carrying abundant pyrite and carbon, which is probably of organic origin. Organised organic bodies in the form of sac-like films of carbon (*Corycium enigmaticum*) have been found in fine-grained greywackes north of Tampere.

Intraformational conglomerates are most common in parts of the succession where volcanic rocks are abundant. The presence of a high proportion of volcanic pebbles indicates that they were derived largely from sources within the basin of deposition, but some conglomerates also carry fragments of granite and migmatite, which appear from isotopic studies by Kouvo to represent early products of Svecofennide plutonism. Sederholm assigned a special significance to certain conglomeratic horizons in the Tampere sequence, regarding them as the basal members of a 'Bothnian' division separated by a stratigraphical break from an underlying 'Svionian' division. Although this interpretation is not now generally accepted, it follows from the occurrence of debris from Svecofennide granitic rocks that the conglomeratic members of the sequence belong to a fairly late stage in depositional history.

The majority of volcanic rocks in the Tampere schist belt are of basaltic or andesitic composition. They include thick groups of pillow-lavas, as well as a remarkable assemblage of basic and intermediate pyroclastics ranging from coarse agglomerates to banded tuffs.

This may be an appropriate place to mention the quartzofeldspathic rocks which are known to Scandinavian geologists as *leptites*. Pale-coloured rocks, usually rather fine-grained and showing a parallel parting, are common components of supracrustal groups in southern Finland and in other parts of the Svecofennide belt. They have generally been regarded as metamorphosed acid volcanics and some, indeed, retain primary textures comparable with those of rhyolites or pyroclastics. In some regions, however, rocks mapped as leptites retain remnants of graded bedding or other sedimentary features and appear to be derived from greywackes, feldspathic sandstones or other psammitic sediments. It seems possible that many leptite groups include metasedimentary material.

The Svecofennian rocks of the Tampere schist belt, as has been mentioned already, pass southwards and south-westward into regions where the grade of metamorphism and the degree of migmatisation are higher. They are represented here by gneisses among which it is possible to recognise derivatives of many of the rock types described. The more argillaceous greywackes are represented by banded biotite-gneisses containing cordierite and garnet, the more sandy types by leptite gneisses composed largely of quartz and feldspar. Basic volcanics

appear as amphibolites and a thick group of these rocks occurring in southern
Finland has been tentatively equated with the main volcanic division of the
Middle Svecofennian. The association as a whole does not, however, correspond
exactly with that seen in the Tampere region. Calcareous rocks, including both
marbles and calc—silicate gneisses, are fairly widespread and the appearance of
these types suggests a southward approach to the original margin of the basin of
deposition. In central Sweden and in the district of Norrland, west of the Gulf of
Bothnia, rocks broadly equivalent to the Finnish Svecofennian reappear in the
western part of the fold-belt: these rocks are often termed the *Svionan*. Around
Stockholm, in an area of advanced migmatisation, greywackes, conglomerates,
leptites (many of them interpreted as acid volcanics) and basic lavas make an
association rather like that of the Tampere schist belt. Sedimentary iron ores
derived from limonitic or sideritic sediments are intercalated in the sequence. In
Norrland a thick series of greywackes is associated with acid and basic volcanics;
the highest members of this series (the *Elvaberg* and *Härnö* Series) include
conglomerates and arkroses as well as greywackes and appear to be uncomformable on the underlying rocks.

2 Tectonics and Metamorphism

The style of structure and metamorphism in the Svecofennide belt is distinctive.
Metamorphism was related to the low-pressure facies series leading to the
regional development of andalusite and cordierite in rocks of pelitic composition. Migmatisation was widespread and led to the formation of numerous
migmatite domes which dominate the structural pattern (Fig. 2.4). These
features suggest that geothermal gradients in the mobile belt were exceptionally
steep.
    In the northern part of the Tampere schist belt, the metamorphic grade is
very low and, as already noted, primary features are well preserved in both
greywackes and volcanics. The dips are generally steep and frequent reversals in
the direction of younging, demonstrated by the orientation of graded bedding
and other depositional or volcanic structures, indicate the occurrence of tight
folds on upright axial planes. Towards the south of the Tampere belt the
metamorphic grade rises and pelitic rocks appear as mica-schists often carrying
porphyroblasts of andalusite and cordierite. Still further south, near the Gulf of
Finland, the rocks become gneissose. Pelitic layers in this southern zone are
generally migmatitic and may contain garnet, cordierite or sillimanite; garnet-
cordierite gneisses of this assemblage have been referred to as kinzigites. The
leptites pass into banded quartzofeldspathic gneisses, sometimes carrying
accessory garnet, cordierite or sillimanite and the basic volcanics into coarse-
grained amphibolites. Schists and gneisses similar to those described here are
widely developed in southern Finland and central Sweden.
    In the region bordering the Gulf of Finland where the metamorphic grade is
high, evidence of a complex tectonic history is apparent. Early fold-systems on
gently inclined axial planes are distorted and partially obliterated by the
characteristic dome-pattern associated with the rise of migmatites at the climax
of the metamorphic history. This region, and especially the coastal tract of

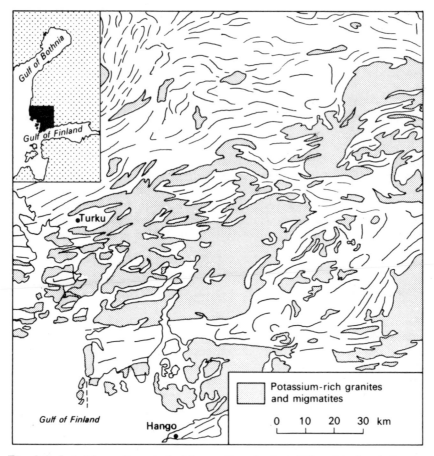

Fig. 2.4. Late-kinematic potash-rich granites of migmatitic origin in south-west
Finland (simplified from a compilation by Härme, 1965)

ice-scoured islands and skerries, provided the terrain from which Sederholm's
concepts of migmatisation were developed and has been described by him in a
succession of *Memoirs* (Sederholm, 1923, 1926, 1930; see also Wegmann and
Kranck, 1931).

The migmatites and granites of southern Finland are classified by Finnish
geologists as late-kinematic. They are microcline-rich rocks ranging from almost
homogeneous granites to mixed gneisses in which a granitic partner is associated
with host-rocks retaining some of their pre-existing structures. The role of
granitic magma in the development of the assemblage has been debated over a
long period. Sederholm at first assigned a major role to incoming magma which
penetrated the country rocks to produce veined gneisses, migmatites, permeation-
gneisses and the like. He came, with time, to lay more stress on the role of
diffuse emanations or *ichors* in migmatisation and to look on the more
homogeneous granitic bodies such as the Hangö granite near Helsinki as

derivatives of *anatectic magmas* generated in the course of migmatisation rather than as representatives of a parent magma. Wegmann and Kranck, in their turn, envisaged the formation of bodies such as the Hangö granite by the activities of diffuse fluids moving through an essentially solid framework. Interchange of material led to granitisation of the host and to the complementary enrichment of the envelope in ferrogmagnesian components such as cordierite – the 'basic front' of some authors. Differences in mobility between the lubricated migmatite bodies and their envelope led to the development of intrusive relationships.

These suggestions could, with little difficulty, be recast in terms of an interpretation involving partial fusion and such an interpretation was indeed favoured by Eskola. The retention of a solid framework during the formation of some of the granites is indicated by the survival of disrupted and recrystallised basic dykes intruded into the country rocks prior to the episode of late-kinematic migmatisation; Sederholm, making use of these bodies to unravel the history of the complexes, provided perhaps the earliest illustration of the value of dykes as time-markers (Fig. 2.5).

3 Svecofennide intrusive suites

The plutonic bodies associated with the Svecofennide belt fall into three main groups: the syn-kinematic, late-kinematic and post-kinematic bodies. All these groups include members that are of granitic character, but they differ in association and mode of emplacement. The *syn-kinematic suites* are, as a whole, characterised by the dominance of sodium and calcium; although intermediate and acid types preponderate, they are always associated with basic or ultrabasic rocks regarded as differentiates of the parent magma. Emplacement of the syn-kinematic suites took place well before the end of the period of orogenic deformation and they often form concordant bodies showing foliations and lineations parallel to those of their country rocks. Considerable regional variations characterise the suite; four provinces have been distinguished in Finland, dominated respectively by granodiorite, trondhjemite, granite and plagioclase-rich hypersthene-granites referred to as charnockitic, while in central Sweden the syn-kinematic suite is partly migmatitic.

The *late-kinematic plutonic rocks* are represented by the migmatites and granites discussed in the previous section; these rocks are potassic and are not associated with basic or ultrabasic differentiates. The *post-kinematic plutonic rocks* form part of the distinctive *rapakivi granite suite* which is distributed across the width of the Svecofennide province and reappears in the corresponding (Ketilidian) province of southern Greenland. It will be more appropriate to consider this suite after the Karelide belt has been dealt with (see p. 35).

4 Svecofennide mineralisation

Mineralisation accompanying the Svecofennide orogeny resulted in the formation of many sulphide deposits in which pyrite is associated with chalcopyrite, pyrrhotite, arsenopyrite, sphalerite, galena or other minerals. These deposits are commonly located in or close to outcrops of Svecofennian basic volcanics or black shales near the borders of syn-kinematic or late-kinematic plutonic bodies. Their formation was sometimes connected with a distinctive magnesia-metasomatism resulting in the enrichment of the country rocks in cordierite, gedrite,

Fig. 2.5. Dykes as time-markers: a metabasaltic dyke (tinted) cuts migmatised conglomeratic schist of the Svecofennian succession and is itself veined by the Rysskär granite (from a photograph illustrating Sederholm's study of migmatisation)

anthophyllite and other Mg-bearing minerals: metasomatism of this kind is seen in the classic area of Orijärvi first described by Eskola where sulphide ores are located at the margin of a syn-kinematic granodiorite. In central Sweden similar ore-deposits are again seen in association with skarns or with rocks enriched in cordierite, gedrite and almandine garnet. In Norrland the important sulphide deposits of the Skelefte region, including that of Boliden, lie on a north-west lineament occupied by volcanics and late-kinematic granites. The ore consists of pyrite, with varying amounts of Cu, Zn, As, Ag and Au. The Kiruna apatite iron-ore body of the same region is emplaced in metamorphosed volcanics at the margin of a syenite.

VII The Karelides

The Karelide belt extending from Karelia north-westward towards the head of the Gulf of Bothnia (Fig. 2.2) flanks the Archaean province of the Baltic shield and is therefore seen in contact with its eastern foreland. It is bordered on the south-west by a large massif of granodiorites and associated rocks belonging to the syn-kinematic suite of the Svecofennides (p. 000). This igneous massif fills the angle between the Svecofennides proper and the Karelides; on its south-eastern side, the easterly trends of the Svecofennides turn sharply northward to unite, in a region of considerable complexity, with the north-westerly trends of the Karelides. A further structural complication is introduced by the occurrence within the Karelides of a median massif in which the Lapland granulites and other rocks of the Belomoride belt are preserved. The boundaries of this massif are major dislocations followed in many places by granitic or basic intrusions.

1 Cover-successions

The *Karelian succession* of supracrustal rocks which rests on the Archaean basement shows considerable lateral and vertical variations which can be related to the history and structure of the belt (Table 2.3). Deposition began early within the mobile belt and was associated with vigorous volcanic activity of several types.

The classic succession of eastern Finland records a marine transgression onto a stable platform and rests directly on the basement; but in Karelia the same succession overlies basin-fillings deposited in early Karelian stages. The basal Sariolian arkoses are patchily developed in eastern Finland as valley-fillings accumulating on a land-surface of low relief. The succeeding marine Jatulian sequence is of transgressive type, with thin oligomict conglomerates and clean-washed, current-bedded, psammites near the base, and finer detrital sediments or dolomites at higher levels. The quartzites of this sequence extend far into Karelia and northern Finland where they overlie older Karelian basin-deposits. Palaeocurrent studies based on the orientation of the false bedding indicate that they were laid down by currents flowing west-north-west from the foreland towards the mobile belt.

Table 2.3 KARELIAN SUCCESSIONS IN THE EAST OF THE BALTIC SHIELD
(based on Kratz, Reitan, Simonen and others)

	Eastern Finland	Karelia	Western Finmark	Kola Peninsula
UPPER	*Kalevian* phyllites and mica-schists of greywacke facies with polymict conglomerates	*Karelian* Basic volcanics, tuffs, slates	*(LAPPONIAN)* sandstones, quartzites, pelites	andesites and basic volcanics, phyllites, dolomites, sandstones, arkoses
	——(*unconformity*)→			——(*unconformity*)→
	Jatulian marine { phyllites, dolomites, Kainuuan quartzites	*Jatulian* slates, dolomites, limestones, quartzites, with basic volcanics	*(LAPPONIAN)* volcanics with carbonate sediments and black pelites; quartzite; basal conglomerate	pelites and amphibolites
	non-marine: *Sariolian* arkoses and conglomerates	*Sariolian* arkoses and conglomerates		
MIDDLE		——(*unconformity*)→		——(*unconformity*)→
		pelites, dolomites, basic, intermediate and acid volcanics		
LOWER		——(*unconformity*)→		*Tundra Series:* Basic and acid volcanics, pelites
		basic and acid volcanics, magnetite quartzites		
	——(*major unconformity*)→	——(*major unconformity*)→	——(*major unconformity*)→	——(*major unconformity*)→
	Basement of gneisses and granites, 2700–2300 m.y.	*Basement of granitic gneisses, 2700–2300 m.y.*	*Basement of gneisses, amphibolites, mica-schists*	*Basement of sillimanite-gneisses, amphibolites, etc.*

In easternFinland the transgressive Jatulian succession is followed by thick pelites of greywacke facies which Wegmann compare with the Alpine Flysch. Little or no volcanic material is presentin this region, but, asTable 2.3 shows, the Karelian of otherlocalities includes a great range of volcanics of orogenic assemblages — spilites, keratophyres, andesites, dacites and basalts are all represented in one area or another.

Serpentines and minor basic intrusions are widely distributed in the volcanic areas. The nickel-bearing differentiated basic intrusives of Pechenga or Petsamo may belong to a late stage of igneous activity. The sediments associated with these assemblages, though predominantly detrital, include dolomites and chert-magnetite iron formations which, though of relatively minor importance, are worth mentioning in view of the abundance of such iron formations in equivalent successions elsewhere. Graphitic shales are minor members of many successions and are occasionally associated with a coaly rock (*schungite*) which contains up to 70 per cent carbon and carries micro-organisms. The occurrence of chemical sediments, as well as the presence of the 'platform' Jatulian sequence, distinguishes the Karelian assemblages from those of the Svecofennides proper. Nevertheless, these assemblages suggest considerable instability and variation in the interior of the Karelides; they are followed in some areas (for example, in northern Finland) by arkoses and conglomerates, which may be of molasse type.

2 Deformation, metamorphism and plutonism

Considerable variations in the style of structures and the intensity of metamorphism are seen in the Karelides. Near the eastern foreland, folding is relatively open and thrusts — perhaps with displacements of the order of 60 km in eastern Finland — carry rocks of the fold-belt eastward. Thrusting is seen also at the margins of the median massif of granulites, where the older rocks appear to override the Karelian successions. The metamorphic grade is low in eastern Finland where much of the Jatulian succession remains almost unaltered. The grade rises towards the junction with the Svecofennides where staurolite and andalusite, or garnet, make their appearance. Migmatisation related to late-kinematic granite-formation is widespread in interior parts of the belt.

Syn-kinematic plutonic bodies are generally comparable with those of the Svecofennides and, like the latter, have yielded isotopic ages of about 1800 m.y. Granodiorites and their associates are most abundant in the western parts of the belt where the metamorphic grade is high. Sulphide-deposits, generally similar to those of the Svecofennides proper, are located in or near some of the ophiolitic basic volcanics and serpentines. Uraninite from the Outokumpu deposit of eastern Finland has given isotopic ages of 1850—1815 m.y., agreeing closely with the dates for Karelide plutonism. It has been suggested that the ore-minerals, which include pyrite and pyrrhotite, with lesser amounts of galena and sphalerite, may have beem mobilised pneumatolytically from the black pelites of the Karelian sequences.

3 The basement in the Karelide orogeny: mantled gneiss domes

In southern Karelia and eastern Finland the basement of the Karelides is distorted to produce a distinctive tectonic pattern. The basement rocks rise into the cover to form numerous steep-sided autochthonous domes on which the Karelian cover-rocks are moulded. The domes are circular or ovoid in cross-section and are often inclined eastward, in conformity with the general easterly vergence of the fold-structures (Fig. 2.6).

It was realised at an early stage of investigation that although the Karelian succession rested unconformably on the basement rocks of the domes, the relations between the two groups were far from simple. In most instances the contacts of basement and cover are strictly conformable with the bedding of the cover successions; the lowest horizons of this succession generally wrap completely round the basement inliers and, where these horizons are conglomeratic, they often contain pebbles of the underlying basement. These features indicate that the cover rocks were deposited after the formation of the basement gneisses. But the gneisses frequently have a foliation that is parallel to their contact with the cover – an arrangement that would demand an extraordinary coincidence if the foliation were older than the deposition of the cover. Moreover, homogeneous granitic rocks sometimes form a capping to the domes and the cover may be invaded by apophyses from these granites. We are therefore faced with the anomaly that the rocks of the basement domes are both older than, and intrusive into, their cover. The solution, propounded by

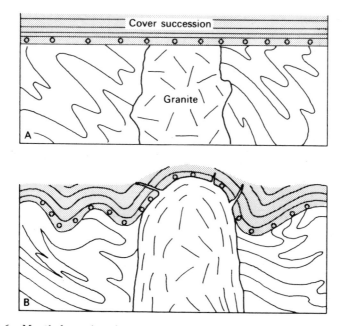

Fig. 2.6. Mantled gneiss domes: diagrammatic sections, based on Eskola, illustrating the structure before (A) and after (B) regeneration of the basement

Sederholm and established in detail by Eskola, depends on the inference that the basement formed during an earlier orogenic cycle was reactivated during the Karelide orogeny, becoming mobile enough to rise into domes, to develop a new concentric foliation and to generate granitic material capable of intruding its cover. The structures so formed were named by Eskola *mantled gneiss domes* (1948). They were formed, in Eskola's view, above the sites of pre-Karelian granitic bodies which were more readily reactivated than the surrounding metamorphic basement rocks. On this interpretation, the rounded shapes of the basement inliers, which control the structural pattern of the overlying Karelian rocks, reflects the shapes of pre-Karelian granitic bodies in the basement.

Eskola's geological reasoning received striking confirmation from isotopic age-determinations carried out by Wetherill, Kouvo, Tilton and Gast (1962). An age-pattern revealed by the dating of several minerals from one sample, and by the use of different methods, reflects the history of reactivation of the basement. The highest ages obtained for feldspar and zircon give approximately the date of the Saamide orogeny. On the other hand, the biotite dates reflect the timing of the Karelide orogeny during which regeneration took place. In a specimen from the Heinävaara mantled gneiss dome, feldspar and zircon remained closed systems with respect to the radioactive and radiogenic components, whereas the biotite recrystallised or became an open system during the Karelide orogeny.

VIII Early Proterozoic Regions of the Ukrainian Massif

The central part of the Ukrainian massif exposed in the lower reaches of the Dnieper forms part of an early Proterozoic mobile belt which has been traced northwards for some hundreds of kilometres beneath the cover of the Russian platform by means of geophysical surveys and borehole-data. This *Krivoi Rog belt* is tentatively linked with the Karelides and, like the latter, gives isotopic ages of about 1800 m.y.

The Krivoi Rog belt of the Dnieper area incorporates Archaean and Katarchaean material which, although partially regenerated, includes several remnants preserving pre-Proterozoic structures and yielding isotopic dates up to 3600 m.y. These remnant blocks are depicted in diagrammatic cross-sections (for example, by Semenenko) as antiformal cores. The early Proterozoic cover-sequence of the belt consists of a volcanic 'metabasite series' followed by a mixed series of volcanics, pelites and banded jaspilitic iron formations. The metamorphism ranges from slate-grade up to pyroxene-gneiss grade. The iron formations of this sequence provide important iron ores in the Ukraine. Magnetic anomalies (the Kursk magnetic anomaly) indicate that they extend northwest along the fold-belt for almost 1000 km beneath the Russian platform; large reserves have been proved by boring in the region south of Kursk where the basement is comparatively near the surface.

For the sake of convenience, we may refer here in passing to the remaining evidence concerning the constitution of the basement beneath the Russian platform. During the last two decades, several hundred boreholes in eastern Europe have penetrated the basement and dating of the samples so obtained has provided a first indication of the Precambrian structure. It will be seen from the map based on Semenenko (Fig. 2.3) that the Archaean and Svecofennide provinces of the Baltic and Ukrainian shields have been traced for some distance under the cover. They are crossed, however, by east—west tracts in which schists, gneisses and granites yield dates of about 1700—1300 m.y. (the Volnian and Gothian-Ovruch folding of Soviet authors). The more southerly of these tracts, which skirts the northern side of the Ukrainian massif, incorporates a thick volcanic and sedimentary series. Its character remains to be clarified — it could be regarded as a portion of the Svecofennide system in which mobility continued after the stabilisation of the branches crossing the Baltic shield or, alternatively, as a wholly post-Svecofennide mobile belt.

IX Post-orogenic Igneous Rocks: the Rapakivi Granites

After the conclusion of the Svecofennide-Karelide orogeny, much of Karelia, Kola, Finland and Sweden attained a state of relative stability which has persisted until the present day. It is perhaps fair to say that the Baltic shield came into existence at this time, for although orogenic mobility was renewed during late Proterozoic and Caledonian cycles around the margins of the present shield, the main Archaean and Svecofennide provinces, with their extensions beneath the Russian platform, remained essentially cratonic.

The final episodes, in which the influence of the Svecofennide patterns of mobility can still be detected, led to the emplacement of granites and associated volcanics in an east—west tract following the Svecofennide belt from Karelia to western Sweden. These post-orogenic intrusions are the *rapakivi granites*. For many years they were regarded as anorogenic and the first radiometric determinations, giving apparent ages of 1640—1610 m.y. appeared to confirm this interpretation. More recent radiometric studies, however, have suggested that the true age of the granites is about 1700 m.y., hardly distinguishable from those for the ending of Svecofennide metamorphism.

The rapakivi granites form large, clearly defined bodies cutting through the folded crystalline complexes of the Svecofennide belts and producing little or no distortion. Small gabbroic or anorthositic intrusions are associated with some of the granites. A set of dolerite dykes is assigned to the same suite and in central Sweden volcanic piles (the *Dala porphyries*, consisting mainly of porphyrites, rhyolitic porphyries and ignimbrites) overlie the roof-region of the granites. These relationships suggest emplacement under a fairly thin cover.

The rapakivi granites have a number of petrological peculiarities which have made them the subject of a notorious controversy. They are highly potassic, containing up to 7 per cent K_2O. Most, though by no means all, contain feldspar ovoids four centimetres or more in length, in which a pink central crystal of potash-feldspar is mantled by a white rim of oligoclase grains. This so-called rapakivi texture is not confined to the rapakivi suite of the Baltic shield, but has been reported sporadically from granite plutons of other regions: from xenoliths in such plutons and from some types of migmatite. Many interpretations have been put forward. Backlund regarded certain rapakivi granites in Sweden as products of granitisation *in situ* of the Jotnian sandstones that overlie them (p. 38). This proposal can be discounted, in view of the strong evidence that the Jotnian sandstones are uncomformable on the granites and the great majority of Swedish and Finnish geologists appear to follow Von Eckerman in regarding the granites as magmatic.

The conditions under which rapakivi textures may be developed during consolidation have been clarified by experimental work on the crystallisation of feldspars. In experimental melts of granitic character a feldspar that crystallises at temperatures above the solvus of the binary system, orthoclase-albite, is a mixed alkali-feldspar containing both K and Na. At temperatures below that at which the cooling curve intersects the solvus, two feldspar species – one potassic and one sodic – begin to appear. Applying these data to the origin of the mantled rapakivi ovoids, Tuttle and Bowen (1958) suggested that the central crystals, which carry some 30 per cent Ab in solid solution, represent the feldspars formed at temperatures above the solvus. As cooling progressed and two feldspar species were formed, they crystallised as mantles around the large feldspars already in existence; those feldspars, mantled by the sodic species, were converted to typical rapakivi ovoids, while the remainder received a coating of potash-feldspars not readily distinguishable from the mixed crystal making the core. The essential feature of the mechanism proposed – that the granites began to consolidate at temperatures above the solvus and completed their crystallisation as subsolvus granites – is entirely consistent with the high-level crustal environment indicated by their field-relations.

X Late Proterozoic Provinces

Much of southern Sweden and Norway is underlain by gneisses and granites whose age-relationships are not entirely clear. Gneiss complexes in south-east Sweden (formerly called Pre-Gothian), which had for many years been regarded as some of the oldest rocks of the country, were found during the late 1950s and 1960s to give isotopic ages in the range 1500–1300 m.y., while gneisses and granites of southern Norway and south-east Sweden consistently gave ages of 1100–900 m.y. Nearly all authors agree that many rocks of the region are polycyclic and date back in origin to Svecofennide or pre-Svecofennide cycles. Some distinguish a 'Gothide' province of mobility at about 1500 m.y. in the south-east and a 'Riphean' or Sveconorwegian province of mobility terminating at about 900 m.y. in the south-west. Others lump the two regions together as a region of prolonged activity in mid-Proterozoic times. The most recent geochronological studies suggest that the south-eastern region may be no more than a part of the Svecofennides in which stabilisation was delayed, whereas the south-western region forms a province in its own right.

1 South-east Sweden

The 'Gothide belt' (Fig. 2.2) is made up principally of large granitic or gneissose complexes between which lie outcrops of quartzites, mica-schists, marbles and basic and acid volcanics. Some of these supracrustal outcrops have been equated with the Jotnian formation (see p. 38), but dating of granitic intrusives in them tells against such a correlation, since these have given apparent ages of as much as 1750 m.y. The southern part of Sweden is traversed by a north-south zone of dislocation some 20 km in breadth, which appears to divide the 'Gothide' terrain giving dates usually over 1300 m.y. from a western region of gneisses giving dates in the vicinity of 1000 m.y. This dislocation-zone truncates the structural pattern of the Svecofennide province and is marked both by many steep shear-zones showing indications of retrogressive metamorphism and by the outcrop of a long narrow syenite dated at >1300 .y.

2 The Riphean province

The zone of dislocation just mentioned marks the eastern boundary of the youngest Precambrian province in the Baltic shield and may perhaps be regarded as the orogenic front of the Riphean belt. The dominant trends within this belt are north-north-westerly and are sharply truncated by the Caledonian front Fig 2.2). In age and geological character the Riphean province has much in common with the Grenville province of eastern Canada (pp. 85–8) and may, inded, represent a continuation of the Grenville belt displaced by continental drift.

The outstanding feature of the province in south-west Scandinavia is the predominance of amphibolite-facies gneisses and other rocks of high meta-morphic grades. Gneisses, migmatites and diffuse granites with abundant pegmatites occupy large areas and pass into complexes in which the meta-morphism is of granulite facies. One such complex, showing the typical features of charnockitic gneisses, is represented by the *arendalites* of the Bamble district of south-east Norway. Many of the gneisses and migmatites appear to represent the regenerated basement of the mobile belt. The large gneiss area of south-west Sweden, for example, is regarded as a horst of basement rocks.

Supracrustal assemblages which may be interpreted as remnants of cover-successions occupy a number of isolated regions and are difficult to correlate. The Kongsberg-Bamble supracrustal groups which occupy the coastal region of south-east Norway, include highly-metamorphosed pelites, psammites, calc-gneisses and amphibolites, intruded by a widespread suite of small basic masses (hyperites) associated with nickel—copper ores or titaniferous iron deposits. All these rocks are partially migmatised and may be components of the basement.

Supracrustal assemblages of lower metamorphic grade which can be fairly confidently assigned to the cover, include the Dalslandian of south-west Sweden and the Telemark series of southern Norway. The *Dalslandian* which is, in places, clearly unconformable on 'Gothian' gneisses, is a group of greywackes, impure psammites, cherts and spilitic lavas, strongly folded, faulted and intruded by late-orogenic granites. It may be a marine 'basin' equivalent of the Jotnian sandstones (p. 000). The *Telemark series*, lying about 100 km west-south-west of Oslo, is some 4000 m in thickness. Its basal parts are mainly volcanic, its middle division mainly psammitic and its highest members include both psammites and volcanics. Above the lowest levels, the grade of metamorphism is low and primary structures such as current-bedding and ripple-marking are well pre-served. Towards the base, however, the metamorphic grade rises and the supracrustal rocks appear to pass by means of a transitional junction into the surrounding granitic gneisses. Intrusive contacts are seen locally and granitic bodies form domes rising into the supracrustals; the relationships of the Telemark series recall those of the Karelian cover-succession overlying the mantled gneiss domes of the Karelide belt (p. 33).

A distinctive igneous suite of the Riphean province is provided by massifs of *anorthosite, hypersthene-bearing gabbro* and intermediate hypersthene-bearing rocks termed *mangerites*. These massifs are situated in regions of granulite facies metamorphism and are themselves often gneissose, garnetiferous and recrystallised, recalling the anorthosites of the Grenville province of Canada (p. 81). They appear at Egersund in south-west Norway and as tectonic slices in the Jotun nappes within the Caledonian belt (Part II).

Late-orogenic granites of the Riphean provinces form a number of discordant plutons invading both the early gneisses and the supracrustal groups. *Pegmatites* are exceptionally numerous and carry a remarkable range of minerals including bismuthinite, molybdenite, fluorite, chrysoberyl, cassiterite, columbite, tantalite, wolframite and topaz.

XI Rocks of the later Proterozoic Craton

The stable craton, established after the termination of the Svecofennide cycle in the central and eastern parts of the Baltic shield, evolved during the later Proterozoic along lines very different from those of the tectonic province discussed in the preceding section. Stabilisation was immediately followed, as we have seen, by predominantly acid igneous activity, during which the Dala porphyries and their associates were erupted and the rapakivi granites were emplaced (pp. 35–6). Sequences of sandstones, arkoses, conglomerates and shales of non-marine facies accumulated in several regions on the eroded surface of the Svecofennide crystalline complexes and rapakivi granites. These *Jotnian Sandstones* are preserved in fault-bounded basins of no great extent, often located not far from the rapakivi centres. They are almost undisturbed and retain primary features – current-bedding, ripple-marking, rain-prints and a prevailing red or brown colouration – which suggest accumulation in piedmont or floodplain environments. Minimum ages of about 1300 m.y. are provided by K–Ar age-determinations on shaly layers. *Post-Jotnian diabases*, members of a widespread dyke suite in the craton, have given minimum ages of 1400–1300 m.y.

The Jotnian sandstones amount to no more than 1000 m in thickness – an inconsiderable amount by comparison with the majority of molasse-type post-orogenic accumulations. Later Proterozoic deposits are also of minor importance on the shield, though they are known to be represented at the base of the Phanerozoic cover on the Russian platform and on the western side of the Ukrainian massif, suggesting that shield and platform areas were differentiated before the end of the Proterozoic.

Cratonic igneous activity in the shield was recorded by the post-Jotnian basic dyke-episode referred to above, by layered nickel-bearing basic bodies, by the accumulation of basalts in Karelia and, more interestingly, by the emplacement of a sequence of alkaline complexes the oldest of which date back to mid-Proterozoic times. These cratonic complexes are represented by syenites and nephaline-syenites dated at between 1800 and 1300 m.y. in Kola and Karelia and by Palaeozoic alkaline ring-complexes. The spectacular Khibine Ţundra complex of nepheline-syenites and allied rocks, with its huge apatite deposits and titano-magnetite ores may be cited as an example of this type. The remarkable uniformity of conditions in the stabilised craton, persisting over many hundreds of millions of years is illustrated by the repeated emplacement of the alkaline complexes from the mid-Proterozoic to the end of the Palaeozoic era.

3

North Atlantic Shield-Areas

I Preliminary: the Structural make-up of the North Atlantic Terrains

The scattered land-areas which lie within and on the borders of the northern Atlantic fall into place as portions of a simpler structure that have been dispersed by continental drift in comparatively recent geological times. The continental fragments which form the bulk of these lands, when reassembled to make allowance for the effects of drift, form three great structural entities (Fig. 3.1):

(a) The Caledonian—Appalachian system of *Palaeozoic mobile belts.*

(b) The more ancient *shield and platform areas* which constituted the stable foreland on the western side of the Caledonian belts.

(c) A fragmentary cover of younger rocks superimposed on these structures and largely concentrated along the *Atlantic margins* of the continents; these margins are fringed in some places by Mesozoic or younger *sedimentary successions* and in others by late Mesozoic—Tertiary *lava-plateaux and associated igneous intrusions* − the components of the classic *North Atlantic or Thulean igneous province.*

The most satisfactory reassembly of continental masses based on a computer controlled matching of the edges of the continental shelf closes the Davis Strait between Arctic Canada and West Greenland, and places southern Greenland close to north-west Scotland. On this reconstruction (Fig. 3.1), the principal Caledonian mobile belt is revealed as a broad continuous tract traversing eastern North America, Britain, eastern and northern Greenland, Spitsbergen and north-western Scandinavia. This belt is mainly of early Palaeozoic age. The western foreland, originally a single area of cratonic continental crust, is represented by three fragments of very unequal size; a small but well-known area clings to the border of the Caledonides in *north-west Britain,* a larger area, partly obscured by ice, underlies *the main part of Greenland* and a still larger area forms *the Canadian shield.*

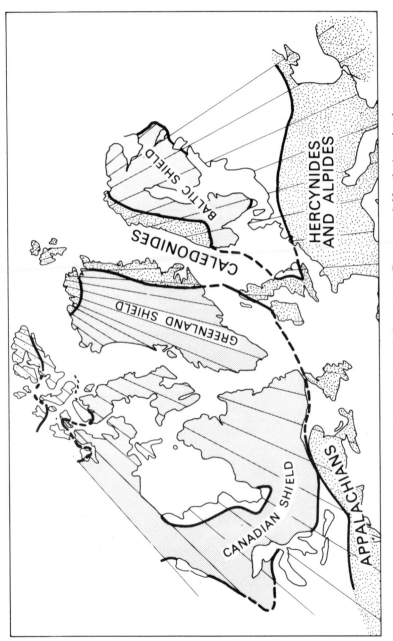

Fig. 3.1. Continental restoration of Greenland, Europe and North America in early Phanerozoic times, showing Precambrian shields, platforms and mobile belts

Set apart from the continental masses mentioned above are the *oceanic regions,* the Davis Strait, the North Atlantic generally, and the oceanic land-areas of Iceland, the Faroes and the smaller islands. In contrast to the ancient terrains along their borders, these units are made up of Mesozoic or younger rocks. They differ fundamentally from the continental areas in structure and history and, as will be shown in Part II, appear to have been formed by processes related to sea-floor spreading.

II The Scottish Shield-Fragment

The fragment of the western shield-area which was stabilised before the beginning of the Caledonian cycle in Britain lies along the western seaboard of Scotland, making the north-west Highlands and many of the offshore islands of the Hebrides (Fig. 3.2 and Table 3.1). It is bounded on the east by the orogenic front of the Caledonides: the Moine thrust-zone. The basement rocks of the shield form a crystalline complex long known as the Lewisian Gneiss but now more often referred to as the *Lewisian complex.* On this rests a cover of late

Table 3.1. GEOLOGICAL SUCCESSION IN THE SCOTTISH SHIELD-FRAGMENT

TERTIARY (mainly Eocene)	plateau-basalts and associated lavas, intruded by differentiated basic and ultrabasic bodies, granites and basic and other dyke rocks
←——————— *(unconformity)* ———————→	
MESOZOIC (mainly Triassic-Jurassic)	thin continental and shallow-water sequences: sandstones, limestones, ironstones, shales
←——————— *(unconformity)* ———————→	
LOWER PALAEOZOIC (Cambrian-L. Ordovician)	transgressive marine succession: ortho-quartzites followed by dolomites: the *Durness succession*
←——————— *(unconformity)* ———————→	
LATE PRECAMBRIAN (c. 900–700 m.y.)	fluviatile, terrestrial and shallow-water assemblages: red arkosic sandstones, conglomerates, shales, greywackes: the *Torridonian Series*
←——————— *(major unconformity)* ———————→	
ARCHAEAN AND EARLY PROTEROZOIC (c. 2900–1700 m.y.)	polycyclic gneisses incorporating products of two major plutonic cycles: the *Scourian complex* (c. 2900–2200 m.y.) and the *Laxfordian complex* (c. 2200–1700 m.y.) together constituting the *Lewisian complex*

Precambrian and early Palaeozoic rocks (which are, at least in part, equivalent to the cover-successions of the Caledonian geosyncline to the east) and piles of Tertiary volcanics overlying a thin Mesozoic succession.

The geology of the small shield-fragment of north-west Scotland has been known in some detail for more than half a century as a result of the

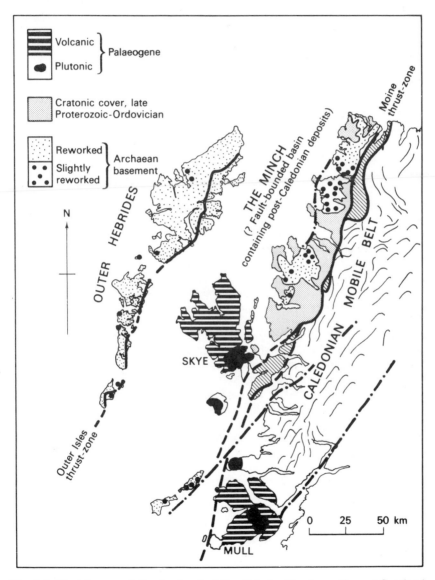

Fig. 3.2. Sketch-map of the foreland of the Caledonides in north-west Scotland showing the distribution of reworked (Laxfordian) and little-modified (Scourian) Archaean gneisses in the Precambrian basement

investigations of Peach, Horne, Clough, Bailey, Harker, Richey and other members of the Geological Survey of Great Britain. The remarkable maps and descriptive memoirs produced by these pioneers gave the first well-documented accounts, not only of certain igneous phenomena displayed in the Tertiary igneous province, but also of a number of basement structures that were later found to be characteristic of polycyclic gneiss provinces in other shield-areas. For this reason, and because we are ourselves familiar with the area, we shall deal with the crystalline Lewisian complex in more detail than its small outcrop area would seem to justify and we shall treat it as an example of the effects of deep-seated regeneration.

III The Lewisian Complex

1 A polycyclic complex

The Lewisian complex may be regarded as a tectonic province stabilised in mid-Proterozoic times, which contains several partly modified remnants of an Archaean complex formed prior to 2200 m.y. The components of the Lewisian may therefore be referred to two principal units:

(a) An older *Scourian complex* of gneisses formed during an Archaean cycle with a time-range of approximately 2900–2200 m.y.

(b) A younger *Laxfordian complex* formed largely by regeneration of Scourian gneisses during a Proterozoic cycle with a time-range of approximately 2200–1700 m.y.

The Scourian rocks are preserved, with relatively slight Laxfordian modifications, in a number of relict massifs enveloped in regenerated rocks which display Laxfordian structural and metamorphic patterns. Remnants of an Archaean supracrustal series occur within the Scourian complex and are preserved in a modified condition within the Laxfordian derivatives of this complex. Post-Scourian supracrustal rocks are of little importance. A large swarm of basic dykes, however, was emplaced in the stabilised Archaean province prior to the onset of the Laxfordian cycle of mobility. This basic dyke swarm often referred to as the 'Scourie dyke' swarm, provides a useful time-marker.

The history of the Lewisian complex is given in a generalised form in Table 3.2 in which an attempt is made to indicate the time-relationships of episodes of many different kinds. Radiometric ages for several phases of igneous and metamorphic activity provide means of linking these episodes with the geological time-scale. The distribution of the principal Scourian remnants and the extent of the regenerated Laxfordian complex that envelops them is shown in Fig. 3.2.

2 The Scourian massifs

The largest surviving Scourian massif is that which makes the central part of the Lewisian outcrop on the western seaboard of the Scottish mainland (Fig. 3.2). Its boundaries run NW–SE parallel to the tectonic grain in the adjacent areas of

Table 3.2. GEOLOGICAL HISTORY OF LEWISIAN COMPLEX

	Volcanic and sedimentary rocks	Intrusive rocks other than granite	Phases of deformation, metamorphism and migmatisation	Granite and pegmatite
LAXFORDIAN c. 2200–1700 m.y.		lamprophyric dykes	metamorphism, migmatisation	pegmatites, granites and pegmatites 1750 m.y.
			repeated deformation, metamorphism of amphibolite or granulite facies	
		SCOURIE DYKE SWARM (tholeiites, etc.)		
SCOURIAN >2900–2200 m.y.	Loch Maree Group?		INVERIAN EPISODE local deformation, metamorphism	pegmatites 2600–2400 m.y.
		diorites, etc.	deformation, metamorphism	granites
			BADCALLIAN EPISODE ≈2800 m.y. gneiss-forming metamorphism, granulite or amphibolite facies	early tonalitic granites?
			interleaving of supracrustal, intrusive and gneissose layers	
PRE-SCOURIAN	Supracrustal groups —— basic/ultrabasic and anorthositic		formation of early gneiss complex	

Laxfordian reworking and its width, measured across this grain, is roughly 55 kilometres. The other massifs are smaller and less clearly defined, ranging down to remnants only a kilometre or so in width, and were at least partially recrystallised in Laxfordian times. Information concerning the general characters of the Scourian complex is supplied not only by these little modified remnants but also by the much larger areas of polycyclic gneisses in which allowance has to be made for the effects of Laxfordian reworking.

From this evidence, it appears that the Scourian complex had reached very high grades of metamorphism throughout its present outcrop well before the end of Archaean times. The predominant rocks are banded gneisses which almost always exhibit a layering or streaking produced by the irregular distribution of ferromagnesian minerals. They contain bands and lenticles of highly metamorphosed metasedimentary rocks and of metamorphosed basic and ultrabasic rocks, all of which appear to have acted as host-rocks during phases of migmatisation and acid veining. The banded gneisses everywhere contain quartz and oligoclase or andesine .and only locally significant amounts of potash feldspar. Their bulk composition is not far from that of intermediate calc-alkaline igneous rocks (Bowes *et al.*, 1971). Potassium, uranium, thorium and rubidium are all low, and Moorbath *et al.* (1969) conclude that these elements were expelled during phases of deep-seated metamorphism which ended at about 2900 m.y.

The rocks produced during these early Scourian phases in the largest Scourian massif appear to have been largely of granulite facies. They are characterised by the presence of hypersthene and clinopyroxene, with garnets in basic rocks, and by a number of peculiarities – the opalescent blue colour of the quartz, the dark greasy appearance of the feldspars and the abundance of antiperthite – which indicate charnockitic affinities. The relationships of metasedimentary, basic and acid components recall those of many migmatite-complexes, in that more basic materials as well as psammitic, pelitic and calcareous gneisses are enclosed by and veined by more acid material in a manner which suggests that the former provided a solid framework acting as host to a mobile acid fraction. Pegmatites are comparatively rare. The early phases of Scourian metamorphism were followed by further episodes of plastic deformation, metamorphism and igneous activity lasting intermittently until about 2200 m.y. and bringing about a partial retrogression to amphibolite facies (Table 3.2).

It may be inferred that partial melting took place during the initial Scourian events at temperatures and pressures sufficient to expel water and components such as uranium and potassium from the complex. Most of the regenerated derivatives of the Scourian gneisses are themselves depleted in these components, although an almost universal retrogression to amphibolite facies indicates an influx of water, and the abundance of potash feldspar in some Laxfordian migmatites suggests a localised infux of potassium and other elements.

The very high metamorphic grades attained during the Scourian cycle and the profound structural modifications which took place during metamorphism make it difficult to determine the original nature of the rocks forming the Scourian complex. Remnants of a metasedimentary assemblage including thin psammites, pelites, marbles, calc-gneisses, graphitic schists and iron-rich gneisses are widely distributed; basic gneisses and pods of serpentine or other ultrabasic rocks are

frequently associated with these metasediments and may represent basic volcanics. These types together form lenses and bands seldom more than a
kilometre or so in thickness. They constitute a supracrustal series which shares
the general Scourian metamorphism and must consequently have been in
existence prior to 2900 m.y. The remaining portions of the Scourian complex
– apart from some late intrusives whose igneous origin is obvious and a number
of earlier anorthosites and basic intrusions – consist of banded granitic gneisses
lacking clear indications of their primary character. Some of these gneisses are
doubtless transformed supracrustals, but it has been suggested by several authors
that they also include derivatives of a gneissose basement on which the main
supracrustal series was deposited in the earliest stages of the Scourian cycle.
Such a pre-Scourian basement would date from a Katarchaean cycle older than
3000 m.y.

3 The post-Scourian dyke swarm

The dyke swarm which traverses the Scourian massifs consists overwhelmingly of
tholeiitic dolerites which are accompanied by dolerites of more basic types and
by ultrabasic dykes composed largely of olivine and pyroxene. A dyke of the
mainland swarm has been dated at 2190 m.y. and an intrusion from the island of
Lewis at 2500 m.y.

The dolerite swarm, as seen on the mainland where it is best-preserved,
maintains a constant west-north-west or north-west trend and has a breadth
perpendicular to this trend of at least 160 km and probably 300 km. Large
dykes occur at the rate of three or four per kilometre and there are, in addition,
numerous dykes less than a metre in width. The dimensions of the swarm,
which extends beyond the limits of the Scottish shield-fragment, show that it is
an entity of regional rather than of local significance. As will be seen later, dyke
swarms emplaced over roughly the same time-span occur in the Archaean
provinces of both Greenland and Canada.

The dykes which traverse the Scourian massifs often retain their megascopic
primary features. Sharp, discordant junctions, chilled margins and narrow
fine-grained dykelets are still clearly recognisable, while sub-ophitic textures are
often preserved and some remnants of the primary minerals survive. Partial
amphibolisation of pyroxenes, the development of garnet rims around ferromagnesian minerals, and clouding of primary feldspars, commonly indicate incom-
plete alteration in dykes crossing the mainland Scourian massifs, while those of
the Outer Hebrides may be wholly recrystallised to hypersthene–clinopyroxene–plagioclase granulites. These modifications in dykes situated in
massifs which escaped extensive Laxfordian regeneration have been variously
interpreted; Dearnley regards them as results of early Laxfordian metamorphism,
whereas O'Hara and others have envisaged reconstitution as a result of emplacement of the dyke swarm at some depth in country rocks which were themselves
relatively hot. The fact that radiometric dates obtained for members of the
swarm overlap with those assigned by Evans to the closing stages of the Scourian
cycle may be significant in this connection.

4 The Laxfordian province

By far the greater part of the Laxfordian complexes exposed in north-west Scotland consists of regenerated rocks which were already gneissose prior to the emplacement of the Scourie dykes. The Laxfordian rocks are therefore poly-cyclic. The only Lewisian rocks which could represent a post-Scourian cover-succession are those of the *Loch Maree Group* (Table 3.2) which form a number of narrow belts sandwiched between masses of regenerated gneiss. This series is made up of thin psammites, pelites, graphitic schists, marbles and calc-schists alternating with thick sheets of basic rocks, all showing a metamorphic imprint of low amphibolite or albite—epidote—amphibolite facies.

The more widespread *polycyclic gneisses* of the Laxfordian complex are generally of amphibolite facies, relatively fine-grained and flaggy-looking where paracrystalline deformation was strong, but often coarse-grained and with isotropic fabrics where recrystallisation outlasted deformation. Large areas, especially in western Harris and Lewis and in the northern and southern parts of the mainland outcrop, appear to have been enriched in potash feldspar and are extensively veined by pegmatites and granitic gneisses carrying abundant potash-feldspar.

Tectonic reworking during the Laxfordian cycle resulted in the development of a new tectonic pattern built up during successive phases of paracrystalline deformation. The style of reworking was characteristically heterogeneous. Remnants of the pre-Laxfordian complex, ranging in size from the central massif of the mainland which is 55 km in breadth, down to pods a few 100 m in breadth, were preserved with relatively minor structural modifications within broader areas in which the complex has been totally reconstituted. These remnant massifs are scattered throughout the Laxfordian province in Scotland and evidently lie well within the original limits of the mobile tract. Most appear to have been defined as structural entities during early phases of Laxfordian deformation and to have controlled the form of the large structures formed during later phases. Those of the Outer Hebrides are located in broad antiformal structures of late Laxfordian age which are separated by synformal tracts characterised by much more thorough deformation and metamorphism. These areas of low deformation escaped complete metamorphic reconstitution during late Laxfordian episodes; metamorphic assemblages of granulite or high amphibolite facies dating from early Laxfordian or Scourian episodes are preserved in some of them and isotopic dates of up to 2600 m.y. have been obtained from them. Tectonic and metamorphic activity were thus closely linked, perhaps because deformation promoted the entry of water into the massive and dry Scourian complex.

On a smaller scale, the inherited characteristics of the gneiss units dictated the style of regeneration. Competent units such as large bodies of igneous origin tended to resist deformation and to retain pre-Laxfordian mineral assemblages. Incompetent units such as thinly banded metasediments were intensely de-formed and their boundaries were often followed by dislocations. Regions of Scourian migmatisation in the island of Harris provided the sites for renewed migmatisation in late Laxfordian times.

On a still smaller scale, the palimpsest structures of pre-Laxfordian age are revealed by such features as the occurrence of agmatites flattened by Laxfordian deformation, of distorted migmatites on which the fabrics formed during successive episodes of deformation are overprinted and by variations in the relationships of the dyke swarm emplaced in the Scourian complex before the onset of the Laxfordian cycle. In areas of low deformation, the intrusive character of these dykes is obvious; they cut discordantly through migmatitic and tectonic structures of Scourian age and retain remnants of primary igneous textures and minerals. Elsewhere, they are recrystallised to pyroxene granulites or more commonly to amphibolites, and folded, boudinaged or disrupted along with the gneiss complex; their discordant relationships are commonly destroyed by distortion of both dykes and gneisses and they are converted to concordant amphibolite bands or pods enclosed in and often veined by the gneisses into which they were originally intruded. The modifications undergone by the post-Scourian dykes provide a rough measure of the intensity of Laxfordian reworking and, where their original orientation can be established, the pattern of structures developed in the dykes may indicate the orientation of Laxfordian stresses during successive phases of deformation.

The inhomogeneity of Laxfordian deformation is reflected in a complex tectonic pattern. The predominant grain is north-west—south-east, but in some regions the structures are moulded around remnant massifs to outline a number of broad domes. Metamorphic boundaries locally transgress the tectonic pattern. Laxfordian migmatites and granite sheets concentrated along the north-eastern margin of the large Scourian massif of the mainland appear to have a tectonic control, but an elongated NNE dome of migmatites and granites in Lewis and Harris is set obliquely across the tectonic grain; this dome may be situated on the site of pre-existing Scourian migmatites. Granite plutons of high-level type are not represented and late-orogenic minor intrusions of all types are rare.

Cutting through Laxfordian provinces along the eastern margin of the Outer Hebrides for a distance of 170 km is a large thrust-zone made up of NNE dislocations, with gentle eastward dips. Extraordinary thicknesses of *pseudo-tachylyte* or 'flinty crush rock' are developed along some of these dislocations and in many places the effects of cataclastic and retrogressive metamorphism are slight. The tectonic significance of the *Outer Isles thrust-zone* is obscure. It lies parallel to the Caledonian orogenic front some 80 km further east, and may therefore be a Caledonian structure, or it may be related to some unrecognised later Proterozoic event.

IV The Greenland Shield

1 The structural make-up of Greenland

The immense Greenland ice-sheet covers most of the shield, with the exception of a discontinuous coastal strip usually from 8 km to 40 km, but in some places up to 160 km in width. Here, the solid geology is magnificently displayed in a bare mountain region penetrated by long steep-walled fjords. Investigations in this difficult terrain have produced results of fundamental importance, especially with regard to migmatisation, to the reactivation of old gneissic complexes and to the differentiation of basic magma. The main elements in the geological structure of Greenland are distinguished in Fig. 3.1 and a somewhat more detailed map is provided in Fig. 3.3.

The main events in the geological history of Greenland fall into three groups as follows:

(a) *Older Precambrian events* >3000–800 m.y., recorded especially in western and southern Greenland. This sequence of events, itself covering several geological cycles, led to the build-up and eventual stabilisation of the *Greenland shield*.

(b) *Caledonian events* 800–400 m.y., recorded in eastern and northern

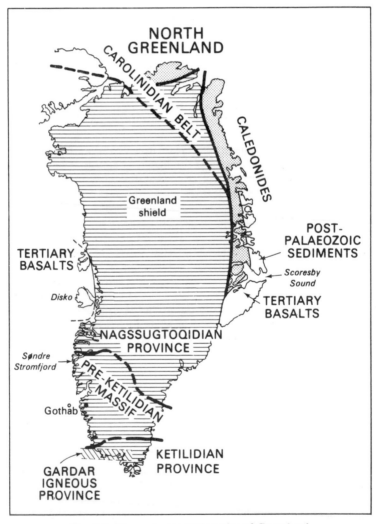

Fig. 3.3. The main tectonic units of Greenland

Greenland where a system of *Caledonian mobile belts* was developed. The cover-successions of these mobile belts are made up of late Precambrian and early Palaeozoic strata and include excellent examples of the deposits of the late Precambrian ice-age.

(c) *Mesozoic and Tertiary events*, recorded by two different types of rocks. *Sedimentary deposits* were laid down during marine incursions at various times in the Mesozoic era and usually lie near the continental margins. *Basic volcanic rocks*, with related *igneous intrusions*, were erupted in coastal regions of eastern and western Greenland during late Cretaceous and Tertiary times. They belong to the Tertiary igneous province of the North Atlantic and include thick plateau basalts and layered gabbros such as the famous Skaergaard complex.

The geological history of Greenland expressed in these terms has a marked similarity to that of western Britain as summarised in Table 3.1. This overall similarity, supported by more detailed examination of such topics as the evolution of the pre-Caledonian shield and the history of Tertiary igneous activity, will have to be taken into account in later discussions of continental displacements.

2 Components of the Precambrian shield

The Precambrian shield of Greenland falls into four tectonic provinces of crystalline or strongly folded rocks, which are overlain in one region by a Precambrian cratonic assemblage. Its five components may be listed as follows (Fig. 3.3):

(a) *The Pre-Ketilidian massif*, of Archaean or Katarchaean age is exposed over a distance of several hundred kilometres on the west coast and reappears from beneath the ice on the east coast. The rocks of this massif are for the most part gneisses. They have locally yielded isotopic dates of over 3700 m.y. and include the oldest rocks so far identified on earth.

(b) *The Nagssugtoqidian belt* occupies a large area in West Greenland, north of the Pre-Ketilidian massif. It forms an early Proterozoic province stabilised at about 1800 m.y. and is made up of highly metamorphosed supracrustal rocks, together with their reactivated crystalline basement.

(c) The *Ketilidian belt* occupies much of southern Greenland on the south side of the Pre-Ketilidian massif and incorporates remnants of a supracrustal cover series, together with its reactivated basement. This province appears to have been stabilised at about 1600 m.y., but the early part of its evolution was roughly coeval with that of the Nagssugtoqides.

(d) *The Carolinidian belt* of north-east Greenland is an orogenic province including folded supracrustal rocks and their basement, which appears to have been formed in later Proterozoic times. Part of this belt is now incorporated in the Caledonides of East Greenland.

(e) *The Gardar assemblage* of south-west Greenland is a cratonic assemblage of sandstones, volcanics and igneous intrusions, some of which have a strongly alkaline cast. The supracrustal Gardar rocks rest unconformably on both Pre-Ketilidian and Ketilidian rocks and the bulk of the assemblage seems to have been formed before about 1200 m.y.

Reference to Fig. 3.3 shows that almost the whole of the shield area which
was to serve as the western foreland of the Caledonides in Greenland had been
stabilised by about 1600 m.y. and that the two principal chronological units –
the Pre-Ketilidian and the Ketilidian-Nagssugtoqidian – compare closely with
the Scourian and Laxfordian units which make up the western foreland of the
Scottish Caledonides. The broad resemblances of age and general character are
complemented by many smaller similarities and a strong case can be made for
regarding the Greenlandic and Scottish forelands as displaced fragments of an
originally continuous shield.

3 The Pre-Ketilidian massif

The Archaean province of Greenland is roughly triangular, narrowing eastward as
a result of the convergence of the Ketilidian and Nagssugtoqidian belts on its
borders (Fig. 3.3). The southern part of the coast of West Greenland reveals a

Plate I. Part of an early dyke-swarm penetrating Pre-Ketilidian gneisses in south-
east Greenland. (Crown copyright)

section through the province some 600 km in length, along which ice-scoured promontories and islands provide an almost ideal terrain for study. The province is made up principally of high-grade gneisses, some of which appear to be polycyclic. Its oldest components have minimum ages of almost 3800 m.y. and constitute a unique assemblage of Katarchaean rocks formed at deep crustal levels. For this reason we shall deal with the massif in a little detail, drawing mainly on the work of the Greenland Geological Survey (G.G.U.).

The tectonic pattern of the Pre-Ketilidian province is complex and certainly polyphase. Over much of the province dome-patterns with no very consistent sense of elongation dominate the regional structure and distort large isoclinal folds of earlier generations. 'Straight belts', in which the dips are generally high and which sometimes contain outcrops of supracrustal groups, define lineaments of northerly trend and appear to represent zones of specialised movement. All these structures are cut by dyke swarms emplaced prior to the Ketilidian-Nagssugtoqidian cycle (Fig. 3.5) as well as by the unconformity at the base of the Ketilidian cover-series and it is clear that the massif had been stabilised by the end of the Archaean. Isotopic dates of over 2000 m.y. for some of the intrusive dykes, and a scatter down to 2200 m.y. for samples from the province suggest that stabilisation was completed at or befoe 2200 m.y.

For descriptive purposes, the rocks of the Pre-Ketilidian massif may be grouped in three assemblages — gneisses, supracrustal rocks and anorthosites and their associates. The gneisses are by far the most widespread types and include the oldest dated components (pp. 55—6). They are predominantly granodioritic in composition but are usually inhomogeneous and incorporate numerous basic layers and occasional remnants of metasediments. Over much of the province they are of amphibolite facies, but they not infrequently show signs of derivation from rocks of granulite facies and unaltered granulie facies, gneisses (enderbitic gneisses) occupy considerable areas.

Supracrustal rocks occupy narrow belts which are interleaved with and folded with the gneisses and which show the same grade of metamorphism as the adjacent gneisses. In the vicinity of Godthåb (Fig. 3.4) the *Malene supracrustals*, which provide typical examples, include basic rocks, occasionally retaining distorted pillow structures and probably largely of volcanic origin, alternating with calcareous, pelitic and psammitic metasediments containing such minerals as diopside, sillimanite, anthophyllite and graphite. The *Isua supracrustals* which lie closer to the ice-cap and appear to belong to an earlier stage (p. 56) include thick silicceous rocks and banded ironstones.

The *anorthosites* and allied rocks occur mainly as distorted bands up to a kilometre in thickness, or as lenses and blocks formed by disruption of such bands. They are invariably metamorphosed and incorporated in the gneiss complex or the associated supracrustals. Where best preserved, as at *Fiskenaesset* in SW Greenland, the anorthositic assemblage is seen to be derived from stratiform igneous bodies, not more than a few kilometres in thickness but extending for many tens of kilometres along the strike. A strong conformable layering indicative of gravitational differentiation is produced by the alternation of anorthosite, gabbro-anorthosite, gabbro, peridotite and pyroxenite. Chromite is a characteristic mineral and is sometimes concentrated in thick layers.

The relationships of the anorthosite assemblage with the enclosing supra-crustals and gneisses show that emplacement of the igneous bodies preceded at

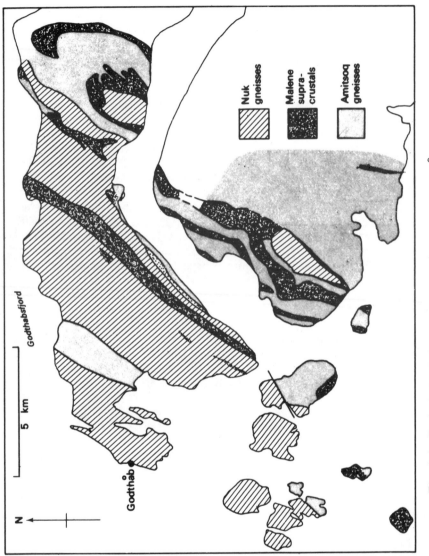

Fig. 3.4. Geological relationships of gneisses in the Godthåb area (based on McGregor 1973)

least one phase of gneiss formation. In the Fiskenaesset region where the gneisses have yielded radiometric ages of up to 3200 m.y., hornblende from an amphibolitic layer in the anorthosite has given a K–Ar date of 3210 m.y. This date provides a minimum age for the high-grade metamorphism affecting both gneisses and anorthosites in the region. Windley has suggested that intrusion of the anorthosites may have taken place as early as 3500 m.y.

The Katarchaean anorthositic bodies of Greenland are of a type which has been recorded from other early Precambrian provinces, notably in Rhodesia, Madagascar and India. Their mode of occurrence and composition – they are rich in calcium and aluminium, carrying a plagioclase rich in anorthite, and are characterised both by chromite and by a high chromium content in the ferromagnesian minerals – distinguish them on the one hand from anorthosites of anorogenic stratiform igneous bodies such as the Bushveld complex and on the other hand from anorthosites of the Adirondack type (p. 82). Windley (1970) has pointed out a resemblance to the anorthositic material thought to be important in the lunar highlands and has suggested that formation of anorthosites of this type may have taken place at early stages in the evolution both of the earth and of the moon.

The relationships of the gneisses, supracrustals and anorthosites in the Godthåb area (Fig. 3.4) throw light on the early history of the earth's crust. A classic field study by McGregor (1973) showed that portions of the gneisses alternating with the Malene supracrustals were derived from granites which had been intruded by a swarm of small basic dykes (the Ameralik dykes) and subsequently converted to gneisses before being interleaved with sheets of supracrustal rocks (Table 3.3). These *Amitsoq gneisses*, which include true

Table 3.3. MAJOR GEOLOGICAL EVENTS IN THE GODTHÅB AREA, WEST GREENLAN]
(based on McGregor, 1973, Black *et al.*, 1972)

8 Repeated deformation and metamorphism, culminating *c.* 2800 m.y. Granulite or amphibolite facies metamorphism associated with depletion of gneisses in mobile elements such as U, Th, K, Rb.

7 Intrusion of tonalites, granodiorites and other calc-alkaline bodies, flooding gneisses and often concentrating at boundaries of supracrustal belts (derivatives of these bodies form the *Nûk gneisses*).

6 Tangential displacements leading to formation of tectonic alternations of units of Amitsoq gneiss and Malene supracrustals.

5 Emplacement of stratiform anorthosites (possibly after 6).

4 Accumulation of *Malene supracrustal groups* including basic lavas, detrital and chemical sediments (possibly contemporaneous with 3).

3 Intrusion of basic dyke swarm (Ameralik dykes)

2 Emplacement of granites, granodiorites etc., deformation and metamorphism converting these to *Amitsoq gneisses* (3700 m.y.)

1 Accumulation of detrital and chemical sediments including iron formations (*Isua formation*, *c.* 3760 m.y.)

porphyritic granites, have an isochron age of not kess than 3700 m.y. and a similar age has been obtained for the Isua supracrustals which include both volcanics and chemical metasediments. The evidence from West Greenland thus establishes the fact that geological processes typical of continental crustal environments − granite formation, dyke intrusion, volcanism and sedimentation − had begun almost 4000 m.y. ago.

In association with the remnants of these very early stages are the Malene supracrustals and their equivalents, the layered anorthosites which are commonly situated within supracrustal belts, and a variety of younger tonalitic and granitic gneisses which appear to be derived from magmatic material flooding into the crust before, or during, the period of widespread high-grade metamorphism (Table 3.3). The concordant interleaving of units derived from many sources has been ascribed to a variety of tectonic processes. Suggestions that the supracrustal units of the interleaved assemblages represent the roots of greenstone belts such as those of the Superior province in Canada (Chapter 4), revealed at a deep level of erosion between regenerated derivatives of the basement on which the greenstones were erupted (Windley and Bridgwater, 1971), imply that the contrasts between Archaean gneiss terrains such asthe Pre-Ketilidian massif and the granite−greenstone belt terrains (pp. 67, 100) are mainly related to depth. Alternative suggestions involving the tectonic interleaving of gneisses, supracrustals and anorthosites by the piling-up of tectonic slices or nappes derived from differing crustal regions implies a more important contrast, possibly related to the distribution of primitive continental and oceanic areas (Bridgwater *et al.*, 1974).

4 The Nagssugtoqidian belt

The tectonic province on the northern flank of the Pre-Ketilidian massif is made largely of gneisses carrying basic and anorthositic remnants not unlike those of the Archaean massif itself. Much of the Nagssugtoqidian belt, indeed, may represent regenerated Archaean gneisses, but it also incorporates supracrustal groups which retain primary features such as current bedding and which probably accumulated in early Proterozoic times on a crystalline Archaean basement. The few radiometric dates so far published suggest that mobility in the belt ceased at or before 1800 m.y.

The tectonic grain is approximately ENE, parallel to the southern boundary of the belt. Some regions of regenerated gneisses exhibit a complex pattern of steep sided whalebacks or domes refolding earlier isoclinal folds and traversed by ENE 'straight belts'. These gneiss complexes include thin belts of supracrustal rocks similar to, and perhaps equivalent to, the Archaean supracrustal groups (p. 53) as well as metamorphosed remnants of the dyke-swarms emplaced in the interval between the formation and the regeneration of the gneisses (see Plate II). The metamorphic grade in the regenerated basement is of amphibolite or granulite facies, even near the orogenic front, and this high grade, with the abundance of basement rocks and the scarcity of late-tectonic granite plutons, suggests that the province reveals a deep section of the Nagssugtoqidian belt. Representatives of the cover succession are most extensively preserved toward

Plate II. Part of the dyke-swarm shown in Plate I, modified by deformation during or after emplacement. (Crown copyright)

the north of the province where mica-schists, quartzites and greenstones form an assemblage of rather lower metamorphic grade.

The southern margin of the province, the *Nagssugtoqidian orogenic front*, crosses Søndre Strømfjord with a general north-easterly trend and has provided a classic illustration of the effects of deep-seated regeneration of gneisses (Noe-Nygaard; Escher, Bridgwater and Watterson, 1973). To the south of the front (Fig. 3.5), massive Pre-Ketilidian gneisses of granulite facies with a general northerly trend are cut by two basic dyke swarms; the earlier and less numerous dykes have an east—west strike and the later and more abundant dykes (the Kangamiut swarm) a NNE strike. As the front is approached, retrogression of the gneisses to amphibolite facies takes place over a 5 km transition zone. The foliation of the gneisses is reorientated to conform with the strike of the front and, more spectacularly, the intrusive dykes become metamorphosed and are rotated to parallelism with the gneisses. Bridgwater, Escher and Watterson record that the east—west dykes show an anticlockwise and the Kangamiut dykes a

Fig. 3.5. Dyke swarms at the southern margin of the Nagssugtoqidian mobile belt in West Greenland. Two intersecting swarms intruding the Pre-Ketilidian gneisses are thinned and rotated in the Nagssugtoqidian belt where the gneissose host-rocks are regenerated (after Bridgwater, Escher and Watterson, 1973)

clockwise rotation, bringing both sets into parallelism with the foliation of the regenerated gneiss. They regard the front as a zone developed on the site of an earlier straight belt along which rocks of the Nagssugtoqidian province have been thrust over the Pre-Ketilidian massif.

5 The Ketilidian belt

The Ketilidian fold-belt which flanks the southern border of the Pre-Ketilidian massif (Fig. 3.3) was identified as a tectonic entity by C. E. Wegmann who recognised the effects of regeneration of the older gneiss complex with its Pre-Ketilidian dyke swarms (See Plate II). It was partly from his work in south-west Greenland that Wegmann derived the concept of an orogenic structure involving a mobile, migmatitic *Unterbau* and a more passively deformed low-grade or non-metamorphic *Oberbau* set out in a classic paper 'Zur Deutung der Migmatite' (1935).

The Ketilidian province was mobile in early Proterozoic times and appears to have been stabilised rather later than the Nagssugtoqides: between 1800 and 1500 m.y. Supracrustal cover-rocks are preserved in a few localities (Fig. 3.5) and a large suite of late-orogenic granites and associated minor intrusives is seen. The regional metamorphism shows considerable variation – portions of both the regenerated basement and the cover are of granulite or high amphibolite facies, whereas some of the higher members of the cover are almost unmetamorphosed. Cordierite is a common mineral in pelitic rocks, suggesting that metamorphism was associated with a rather steep geothermal gradient.

The *Ketilidian cover-succession* is seen in an undisturbed condition resting on gneisses of the Pre-Ketilidian massif north of the Ketilidian front. An important dyke swarm (the Kuanitic dykes) cut the basement but do not enter the cover. In this cratonic region, details of the succession are well preserved and it has been shown that the earliest formations accumulated in several shallow basins which later coalesced to receive the main sequence. Clastic sediments and basic volcanics form the bulk of the sequence, but more than one member includes dolomites and cherts, a few of which carry well preserved organic remains. Iron-rich beds carrying magnetite and hematite recall the banded iron-formations of other comparable successions. Within the Ketilidian belt, the extent of the corresponding succession has not been established. Thick psammites and semi-pelites and major groups of acid to intermediate volcanics occur, but may be older than the Ketilidian succession proper.

In the region north of Ivigtut, the cover is clearly unconformable on Pre-Ketilidian gneisses, and Kuanitic dykes in these gneisses show little deformation. Progressive structural changes take place towards the Ketilidian front. The cover-succession becomes folded and its metamorphic grade rises. Its unconformable base is replaced by a zone of schistosity and in some places is invaded by granitic material. The basement gneisses are traversed by shear-zones or more thoroughly regenerated, and the Kuanitic dykes are progressively distorted and recrystallised, finally passing into lines of concordant amphibolite pods. Both basement and cover pass into areas of migmatisation such as that which constitutes the 'Julianehåb granite'.

A remarkable array of late-orogenic granites and associated intrusives invades the Ketilidian belt, providing a contrast both with the Nagssugtoqidian belt of Greenland and with the Laxfordian complex of Scotland. The suite includes granites of various types, among which are rapakivi granites closely resembling the roughly contemporary rapakivi granites of the Baltic shield; appinitic bodies; lamprophyric and basic dykes; and high-level acid minor intrusions and lavas. It has much in common with the late-orogenic igneous suite of the British Caledonides. The term *Sanerutian*, which has been used in various senses, is now generally applied to the late orogenic events connected with the emplacement of this array.

6 The Gardar assemblage

The rocks grouped under the name Gardar in southern Greenland include a post-orogenic supracrustal series, the *Gardar formation,* which rests on a peneplained surface of crystalline rocks in the Ilimaussaq region, a somewhat younger set of alkaline intrusions and a widespread basic dyke swarm. The alkaline intrusions have given radiometric dates of 1250–1000 m.y., and certain early dolerites dates of 1400 m.y.; the Gardar formation is bracketed between the older of these dates and the dates of about 1600 m.y. given for stabilisation of the Ketilidian belt.

The *Gardar formation* is preserved mainly in a fault-bounded block near Ilimaussaq. It reaches a maximum thickness of about 3 km and consists principally of sandstones (often red in colour and sometimes of eolian origin), conglomerates, basalts and pyroclastics. It has much in common with the Jotnian sandstone (p. 38) and has some of the features of a molasse.

The *Gardar intrusions* are concentrated in the region north and west of Julianehåb (Fig. 3.3) where they appear to be emplaced at the intersections of several ENE zones of faulting and shattering with a block bounded by two major wrench faults of WNW trend. The localisation of the alkaline province and its association with faulting are noteworthy. Among the principal intrusions are those of Ilimaussaq, Nunarssuit and Ivigtut with which are associated the cryolite deposits of Ivigtut – now worked-out – and a number of small deposits rich in thorium and uranium. The plutonic rocks include a wide variety of syenites, some nepheline-bearing and others sodalite-bearing, with alkali-granites, alkali-gabbros and gabbros. There are at least half a dozen plutonic centres, the largest of which is 45 km in diameter; individual intrusions form steep sided plugs or ring-structures, some of which show spectacular igneous layering attributed to gravitational differentiation *in situ.*

Gardar dykes are most varied and abundant near the plutonic centres, but regional swarms, composed principally of dolerite or olivine-dolerite extend throughout south Greenland. Within the alkaline province, there are also alkaline, acid and intermediate dykes some of which are giants measuring 500 m across. An extraordinary feature is the presence of crowds of anorthosite xenoliths in dykes occurring within a zone extending for 200 km westward from Narssarsuak.

V The Carolinidian Belt

In the north-west and north-east corners of Greenland, Precambrian cratonic cover-rocks can be seen resting unconformably on rocks of the crystalline shield. In Peary Land and East Greenland, the shield is bordered by branches of the Caledonian system of mobile belts in which even the youngest Precambrian supracrustal formations are deformed. Partially obliterated by these Caledonian zones is a somewhat older fold-belt which appears to be intermediate in age between the (undated) crystalline rocks of the main shield and the Caledonides. The orogenic activity in this *Carolinidian belt* may have been of Grenville age; it could be considerably older or, alternatively, it could represent an early phase of the Caledonian orogenic cycle. It is dealt with here largely as a matter of convenience.

The Carolinidian belt extends northwards for several hundred kilometres from Dronning Louise Land towards Peary Land and Ellesmere Island. Its outcrop is bounded on the east by the orogenic front of the Caledonides and on the west by the margin of the ice-sheet. Within the fold-belt is a thick series of detrital sediments totalling some 6 km, which constitutes the *Thule Group* of Koch and of Haller. The lower portion of the group, mainly pelitic or semipelitic, is restricted to the fold-belt. The upper portion, consisting largely of sandstones, spreads onto the foreland region to the west and south and is thought to be equivalent to sandstones resting unconformably on crystalline rocks in northern and north-west Greenland. Basaltic dykes and sills invade the Thule Group throughout north-west Greenland. The effects of orogenic folding and metamorphism are seen within the main Carolinidian belt and decrease westward towards its foreland.

4

Precambrian of the North American Craton

I Preliminary: The Structural Make-up of North America

The North American continent is made up of three great structural entities (Fig. 4.1). These are the *North American craton* which has the form of a large triangular block with its apex towards the south; and the Phanerozoic fold-belts of the *Appalachian* and *Cordilleran systems* which flank the craton to the south-east and south-west respectively. The Appalachian fold-belt, largely of Palaeozoic age, was stabilised very early in the Mesozoic era and is partly covered by a blanket of younger sediments forming the coastal plain and continental shelf of the Atlantic margin. The Cordilleran belt has been active since late Precambrian times and has yet to be stabilised. A zone of seismic activity at the Pacific margin of the continent is continued in the island arc systems of the Aleutians and central America which link North America with Asia and South America.

The North American craton, the core of the present continent, is underlain by Precambrian rocks which have shown only limited mobility in Phanerozoic times. These Precambrian rocks crop out widely in the north to form the *Canadian shield*, the main subject of this chapter, and are revealed further south in a few broad upwarps and in up-faulted blocks entangled in the Cordilleran system; but over the greater part of the interior lowlands of the United States, the Precambrian is hidden beneath a cratonic cover of Phanerozoic sediments.

II The Canadian Shield

The Canadian shield forms the largest single region of Precambrian rocks in the world. The needs of an expanding mineral industry have stimulated geological

Fig. 4.1. The main tectonic units of North America

investigation and the greater part of the shield has been mapped at any rate on a reconnaissance basis. The principal mining areas have been studied in great detail and regional geological, tectonic, gravity and aeromagnetic maps have been published in recent years.

Structural provinces. A broad subdivision of the Precambrian shield was achieved before the advent of isotopic dating by the recognition of a number of *structural provinces*, characterised by a certain consistency of structural pattern and style of metamorphism and mineralisation, as well as by the abundance of particular lithological assemblages. This concept, which has since proved to be applicable to nearly all shield areas, arose from a long train of field investigations, beginning with the reconnaissance of Logan and Sterry Hunt in the mid-nineteenth century and developing through the early decades of the twentieth century with the work of such geologists as Stockwell, Wilson, Quirke,

Collins, Derry and Gill. Its fundamental soundness has been confirmed in the last couple of decades by regional geophysical surveys and by the results of isotopic dating.

An aeromagnetic map covering a large part of the shield, published by the Geological Survey of Canada in 1968, displays a pattern of anomalies which, naturally, reflects the distribution in depth of rocks of differing magnetic properties. Several conspicuous lineaments marking discontinuities in the pattern, or separating patterns of different styles correspond with the borders of structural provinces. *Bouguer gravity anomalies* are also located along some provincial borders.

A systematic programme of isotopic dating, mainly by K–Ar methods, embarked on by the Geological Survey of Canada in the late nineteen-fifties, has resulted in the publication of hundreds of apparent ages for metamorphic and granitic rocks from the shield. These data indicate that the rocks of each province were stabilised and cooled over well-defined time intervals which differ from one province to the next. The shield may therefore be regarded as a patchwork of provinces each of which is a portion of a Precambrian mobile region formed during one group of geological cycles.

The broad features of the Canadian shield and of its extension beneath the central lowlands of the United States are presented in the diagrammatic map of Fig. 4.2, Table 4.1 and the histogram of Fig. 4.3. It will be seen from the latter that the provinces of the shield were tabilised during three main periods which are interpreted as the closing stages of three cycles of orogenic mobility. Canadian geologists have devised a twofold terminology in which one series of names is used todenote the climaxes of successive cycles and a second series of time-stratigraphic terms is used to denote the 'eras' or time-periods occupied by each successive cycle. All these terms are summarised in Table 4.1, but the second series, which does not seem to have caught on outside Canada, is not used in this chapter.

The main provinces are few and large and it is clear from their structural relationships that they represent portions of originally still more extensive complexes. For example, the Slave and Nain provinces are small massifs of little modified rocks of Kenoran age enclosed in rocks affected by younger cycles, many of which appear to be polycyclic. It seems probable that the *Kenoran orogeny* affected all the regions between the Slave, Nain and Superior provinces, as well as a large area to the south-west of the exposed shield. No limit can at present be set to the original extent of the Kenoran tectonic and metamorphic processes.

The *Hudsonian mobile zones*, on the other hand, show clearly-defined fronts against the Superior province, which evidently constituted a stable block during the Hudsonian cycle. This cratonic block must have been relatively small, for there are indications of Hudsonian folding and metamorphism to the north-east, north and south. Some portions of the Churchill, Southern and Nain provinces – all of which show the effects of Hudsonian mobility – have yielded K–Ar ages which range down to about 1300 m.y. These regions are assigned by Stockwell to a later Elsonian cycle (Table 4.1). Whether these dates indicate events equivalent in status to the Hudsonian and Kenoran cycles seems open to question, as will be seen later. The *Grenville belt* appears to have been flanked

Table 4.1. CYCLES AND PROVINCES OF THE CANADIAN SHIELD
(based mainly on Stockwell, 1964, 1968)

Time m.y.		Orogeny	Provinces	Era
600				HADRYNIAN
1000		Grenville (climax 1000–900 m.y.) (= Llano)	Grenville	HELIKIAN
1500	PROTEROZOIC	[Elsonian (climax 1500–1300 m.y.)]	[Parts of Nain, Southern]	
2000		Hudsonian (climax 1800–1650 m.y.) (= Penokean)	Churchill, Bear, Southern, part of Nain	APHEBIAN
2500				
3000	ARCHAEAN	Kenoran (climax 2600–2400 m.y.) (= Algoman)	Superior, Slave, part of Nain	ARCHAEAN

on the north-west by a stable block, little narrower than the present Canadian shield. It is itself a belt of moderate width which can be proved along its strike as far south as the Mexican border.

III Extensions of the Shield

At the southern border of the Canadian shield the Precambrian basement descends beneath the Phanerozoic cover to depths averaging a kilometre or so below the surface. Basement inliers are small and widely separated, except for a string of inliers upfaulted in the frontal region of the western Cordilleras, but the information they provide is supplemented by material from nearly six thousand boreholes drilled to basement, the majority lying west of the Mississippi in areas of interest to the oil industry. From this material, and from the results of geophysical investigations, preliminary maps of the basement underlying the greater part of the North American craton have been prepared (Muehlberger and others, 1967).

Rocks dating from about 2500 m.y., broadly equivalent to those of the Superior province, have been identified over considerable areas in the region west of Lake Superior, where they represent a direct extension of the province, and in at least one massif further south-west, underlying much of Wyoming (Fig. 4.2). Provinces involved in Hudsonian (and Elsonian) orogeny enclose the Archaean massifs and emerge near the Pacific coast over a broad front near the Gulf of California. A Grenville province, directly continuing that of the Canadian shield, embraces the eastern and southern border of the craton, providing the basement of the Appalachian mobile belt. A point of interest which will be returned to later (p. 82) is the preservation in some of the southern parts of the basement of a little-disturbed cratonic cover of later Proterozoic supracrustal rocks.

IV The Archaean Provinces

1 Distribution of Archaean rocks

Although very large areas of early Precambrian rocks have remained essentially unmodified in the North American craton since the ending of Archaean times, very few rocks yielding ages of over 3000 m.y. have been located. The scarcity of Katarchaean ages is probably to be attributed to the effectiveness of Kenoran orogenic activity in reworking pre-existing material (cf. p. 70). A remnant of very old material which appears to have escaped regeneration occurs in south-western Minnesota where granite-gneisses which form a basement to mid-Precambrian cover-rocks have given a whole-rock Rb–Sr isochron age of about 3500 m.y., an age confirmed by U–Pb dating of zircons.

The *Superior province*, with its south-western extension into U.S.A., provides one of the largest massifs of Archaean rocks in the world and as parts of it are richly mineralised it has been explored in considerable detail. As has been noted, it seems reasonable to envisage the existence of an Archaean region of mobile

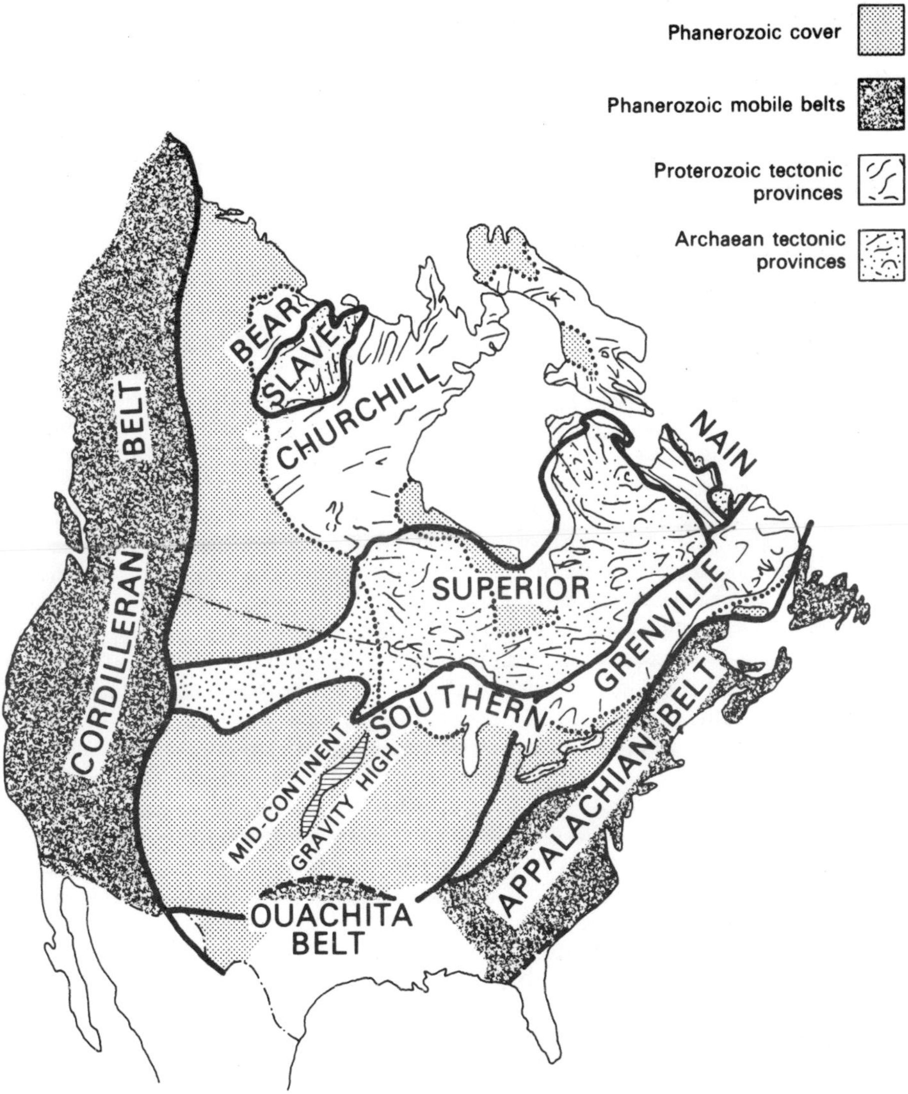

Phanerozoic cover

Phanerozoic mobile belts

Proterozoic tectonic
provinces

Archaean tectonic
provinces

Fig. 4.2. Major features of the Precambrian basement of North America

crust at least half the present size of the North American continent, with no
known boundaries against older cratonic areas. The present Archaean provinces
are simply remnants of this immense region.

Two principal assemblages make up the greater part of the Archaean
provinces. A supracrustal assemblage characterised by abundant basic volcanics
occupies numerous elongated tracts up to several hundred kilometres in length,
which are often referred to as *greenstone belts*. An assemblage of *granites,
gneisses and migmatites* appears in the broader and more irregular areas between

the greenstone belts and, in a very general way, occupies the lower structural levels. The greenstone belts, trending roughly east and west in the Superior province and nearly north and south in the Slave province, define the tectonic grain of the Archaean complexes. They carry deposits of gold and associated metals which have provided some of the main resources of Canada's mining industry. *Post-Kenoran supracrustal rocks* lying undisturbed on the folded and metamorphosed Archaean rocks are preserved in a number of areas, especially near the borders of the provinces. These rocks will be dealt with in connection with the Proterozoic cycles of mobility.

2 The greenstone belts

The supracrustal assemblage of the greenstone belts is of a kind which is widely represented in Archaean provinces of Africa, Western Australia and India where it is often associated with gold mineralisation. The typical greenstone belts contain both metavolcanics and metasediments, though the proportions vary greatly. The volcanics are preponderantly basic, often pillowy, but contain important groups of andesitic or more acid lavas and pyroclastics. Chert beds occur in association with the volcanics and with serpentines and intrusive basic bodies. The sediments are detrital and largely of greywacke facies, the whole assemblage suggesting accumulation in unstable basins.

Although a generalised succession in which volcanics predominate towards the base and metasediments towards the top seems to be common to several greenstone belts, both in the Superior and in the Slave province, it is doubtful whether there is a basis for direct stratigraphical correlation from one belt to another. Evidence from isotopic dating suggests that the greenstone belts vary considerably in age. Thus, a whole-rock Rb–Sr isochron for volcanics of the Yellowknife belt in the Slave province gave an age of 2625 m.y. (Green *et al.*, 1968), and dates of about 2700 m.y. have been obtained from the Rainy Lake area, whereas isochrons for volcanics in the Kirkland Lake area of Ontario gave ages of about 2370 m.y. The occurrence of thick wedges of polymict conglomerate containing abundant volcanic debris suggests derivation from local sources and indicates that consistency of succession beyond the limits of one belt is not to be looked for. It seems probable that individual zones of vulcanity evolved independently, building up volcanic piles which then contributed erosional debris to adjacent sedimentary basins; a similar evolution characterises many island arc systems with which the greenstone belts have been compared.

In the early days of exploration, a twofold division into a lower volcanic group and an upper, uncomformable, sedimentary group was established in the Kirkland Lake area, in which some of the principal mining centres in eastern Ontario were situated. The names *Keewatin* and *Timiskaming* were given, with a stratigraphical meaning, to these groups. As reconnaissance was extended, these names came to have little more than a lithological significance, volcanic groups being assigned to the Keewatin, and sedimentary to the Timiskaming. Their stratigraphical value is now doubtful.

According to Goodwin (e.g. 1968) the typical succession of the greenstone belts records three stages of igneous activity. The first is characterised by many kilometres of tholeiitic basalts, spilitic basalts and andesites with related intrusions erupted over quite wide areas. The second is marked by the incoming of andesitic, and finally of acid, pyroclastics and lavas, forming more restricted piles up to several kilometres in aggregate thickness. The third is characterised by only occasional volcanic outbursts interrupting the accumulation of sedimentary rocks, formed largely by erosion of the earlier volcanics. The majority of the basic lavas appear to have been submarine. The later acid volcanics, with their abundant pyroclastics, are concentrated around centres which may have been built up into subaerial volcanoes. The basic→acidic trend, occasionally repeated in a second cycle, is interpreted as a result of differentiation of a calc-alkaline parent magma.

The metasediments of the greenstone belts contain a considerable proportion of badly sorted clastic debris, along with greywackes, shales and minor amounts of chemical sediment. The most important of these chemical sediments are iron-rich cherts which commonly occur close to volcanic formations and which are thought to derive much of their substance from volcanic emissions. Coarse polymict conglomerates, often carrying abundant detritus from volcanic rocks, tend to occur in wedge-shaped masses up to several hundred metres in thickness; they appear to have been derived from mountainous source-areas close to the region of deposition. The badly sorted clastic and detrital sediments which alternate with, or succeed, the volcanics exhibit graded bedding and other structures characteristic of turbidites, suggesting rapid deposition in unstable basins. Formations which occur at some distance from the volcanics are composed predominantly of well-bedded quartzo-feldspathic psammites, pelites and greywackes. These formations are probably as extensive as the assemblages rich in volcanics, though they have received less attention. Goodwin considers that terrains of crystalline 'granitic' rocks must have been exposed to erosion in the hinterlands between active volcanic belts: it is worth noting that even the conglomerates of the greenstone belts themselves contain some granitic debris.

3 Structure and metamorphism

The rocks of the greenstone belts show complex patterns of folds on axial planes which are predominantly steep. The belts themselves define the tectonic trend in the Superior and Slave provinces and are generally regarded as synformal tracts; the intervening regions of granitic rocks with metasediments form broad zones of antiformal character.

The rocks of the greenstone belts often show a very low grade of metamorphism, the characteristic mineral assemblages being of greenschist facies. Primary structures are preserved in lavas, pyroclastics and turbidites, and in some localities even the microscopic textural details and the primary minerals can still be recognised. In the predominantly granitic tracts on the other hand, any supracrustal rocks tend to be of higher metamorphic grade. Cordierite and andalusite are common in the pelitic rocks in some of these tracts, suggesting the operation of regional metamorphism with a high geothermal gradient. Towards the north-east and west of the Superior province, the metamorphic grade rises and rocks of granulite facies cover considerable areas. Rocks of amphibolite facies, closely mixed with gneisses and granites, form much of the Wyoming massif.

4 Granitic rocks

The greenstone belts of the Superior and Slave provinces are separated by broader tracts in which granite-gneisses, migmatites and granites predominate. Metasediments and metavolcanics are not absent from these tracts, but differ from those of the well-defined greenstone belts in showing a high grade of metamorphism and in being not infrequently migmatised.

At the borders of the greenstone belts, fairly homogeneous granodiorite, or granite, is frequently seen in intrusive contact with the supracrustal rocks and similar rocks may penetrate even the younger formations in the interior parts of the belts. These intrusive bodies have been assigned to several groups (for example, the Laurentian and the Algoman granites of the Superior province). They have yielded K—Ar ages which are generally in the range 2700—2400 m.y. and can be regarded as products of plutonic activity during the Kenoran cycle.

The gneissose granodiorites, migmatites and gneisses which occupy the greater part of the 'granitic' tracts are of rather more doubtful origin. These rocks yield K—Ar ages in the same range as the intrusive granites and have undoubtedly been strongly affected by Kenoran plutonic processes. Their structural position, as tracts separating the apparently synformal supracrustal belts, suggests that they may incorporate material from the basement on which supracrustals accumulated. We have already noted that granitic sourcelands were exposed to erosion during the deposition of some of the sedimentary formations and we may tentatively regard the tracts of gneiss and migmatite as including a regenerated basement of Katarchaean age.

5 Mineralisation

An important sulphide mineralisation characterised by gold-quartz lodes sometimes associated with pyrite, arsenopyrite, pyrrhotite, sphalerite or galena is seen in both the Superior and Slave provinces. Gold provided the basis of the mining industry in such localities as Kirkland Lake, Porcupine, Noranda and Yellowknife. The deposits, which are located mainly along faults or shear-zones, are concentrated in the low-grade rocks of the greenstone belts especially near centres of acid vulcanicity. Their emplacement dates from late stages of the Kenoran orogeny; their source has been considered by many authors to be granites of the Kenoran cycle, but may more probably have been the volcanics of the greenstone belts themselves.

6 Stabilisation of the Archaean provinces

By 2400 m.y., Kenoran orogenic activity was greatly reduced and a very widespread stabilisation and cooling of the crust appears to have taken place.

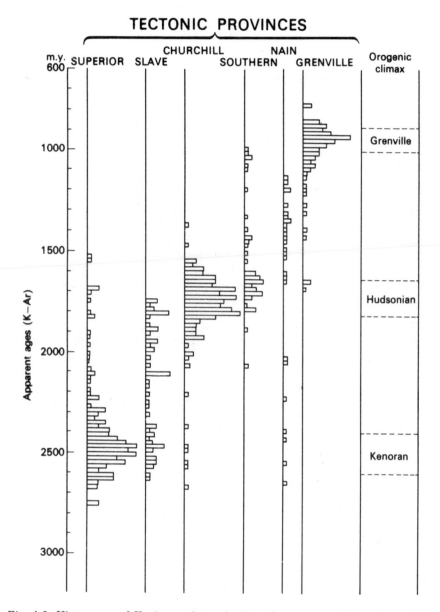

Fig. 4.3. Histograms of K—Ar age-determinations for metamorphic and granitic rocks from six tectonic provinces in the Canadian shield (based on Stockwell, 1968)

These events are indicated by an abrupt fall in the number of K—Ar dates obtained from rocks of the Superior and Slave provinces (Fig. 4.3) and by the fact that several undisturbed Proterozoic formations giving apparent ages up to 2300 m.y. rest on the folded and metamorphosed rocks of the provinces.

7 Basic igneous activity of late Archaean times

After the phase of cooling and stabilisation, the Archaean provinces of North America were invaded by very extensive swarms of basic dykes and sills. In the Slave province and in the Ungava peninsula in the north of the Superior province, dyke swarms emplaced at this time have a general east-north-east trend. The swarms lie roughly on a great circle and if, as seems probable, they represent remnants of an originally continuous set, its length must have been not less than 2500 km. Payne and others (1964) have pointed out that if the great circle representing the Canadian swarms is projected eastward on a continental reconstruction allowing for the effects of Phanerozoic continental drift, it passes close to the dyke swarms of similar age in the Pre-Ketilidian complexes of Greenland and the Scourian complexes of Scotland.

A plutonic basic body emplaced in the Archaean massif of Wyoming and Montana is the huge *Stillwater complex* of the Beartooth Mountains which has the form of a tilted sheet more than 5 km in thickness. This complex shows a well-developed layering attributed to gravity-settling of minerals during crystallisation and includes peridotites with chromitite bands, gabbros, norites and anorthosites. The rocks have suffered no significant metamorphism, but have been tilted into an almost vertical position. Considerable interest has been aroused in recent years by the publication of K—Ar and Rb—Sr ages of up to 3200 m.y. for rocks of the complex and its aureole. The most recent determinations of Fenton and Faure (1969) however, have yielded an isochron age of 2450 ± 210 m.y., a result which would be consistent with emplacement soon after the end of the Archaean cycle of mobility in the Wyoming massif. The time-span of the basic igneous activity which followed the phase of stabilisation ranges from about 2450 m.y. down to 2150 m.y. (the age of Nipissing diabase sills intruding the Huronian, p. 74) and possibly to about 1900 m.y.

V Early Proterozoic Provinces

1 The Hudsonian system of mobile belts

The Hudsonian mobile belts of the Canadian shield follow an arcuate course north-eastward through Alberta and Manitoba to the Arctic islands and thence south-eastward through Labrador. The belt on the western side of Hudson Bay between the Slave and Superior provinces is almost 1000 km in breadth. Another branch traverses the Grenville province (where it has been modified by later Proterozoic activity) and proceeds via the Southern province in the vicinity of the Great Lakes south-westward beneath the Phanerozoic cover to the Rocky

Mountains (Fig. 4.2). Clearly defined orogenic fronts mark the boundaries of the Hudsonian belts against the older cratons and are characterised not only by rapid decreases in the intensity of deformation and metamorphism but also, in some areas, by changes of facies and thickness in the Proterozoic supracrustal successions. Long-continued contrasts in the behaviour of mobile and stable regions are clearly indicated. The termination of the Hudsonian cycle at about 1750 m.y. (or, where Elsonian activity was recorded, at about 1400 m.y.) was followed by the welding of the stabilised belts onto the cratons to form a much larger stable mass. The relationships between the crustal movements involved and those characteristic of later eras are discussed on p. 179.

The Churchill province and other areas affected by the Hudsonian cycle are occupied partly by supracrustal rocks of early Proterozoic age and partly by regenerated Archaean complexes equivalent to those described in the previous pages. The cover-successions are of considerable interest on account of the presence of thick banded iron formations which represent some of the principal sources of iron-ore in North America. We shall deal with three principal assemblages (Fig. 4.4) — the *Huronian* of the type-area in eastern Ontario, which fringes the Archaean craton; the *Animikie* of Lake Superior, which is seen both

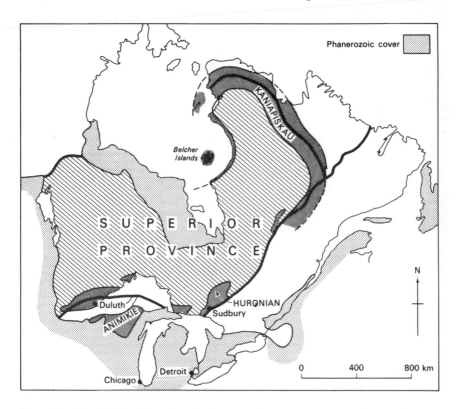

Fig. 4.4. The distribution of early Proterozoic cover-sequences (the Huronian and other groups) in relation to the Superior province

on the southern fringe of the craton and in the mobile belt; and the *Kaniapiskau* in and on the margin of the Labrador trough.

2 The Huronian succession

The classic Huronian succession, named by Logan and Hunt in the middle of the nineteenth century, extends through a belt some 300 km long running north-eastward from Lake Huron along the eastern margin of the Superior province. Towards the west of the outcrop. it is only gently folded, but the amount of tectonic disturbance and the grade of metamorphism increase sharply near the Grenville front. Within the Grenville province, Huronian rocks are strongly-migmatised and few details of their stratigraphy are known. An isochron age for the Gowganda formation of 2280 ± 87 m.y. and a date of 2150 m.y. for sills of Nipissing diabase show that the Huronian is well over 2000 m.y. in age.

The Huronian of the Blind River area to the north of Lake Huron, as befits a foreland succession, is no more than a few kilometres in thickness and consists largely of current-bedded clastic sediments. Two principal divisions were recognised by Collins, the lower *Bruce* division being restricted to an east-west trough crossing northern Lake Huron, and the overlying *Cobalt* division overstepping onto the basement. Towards the east, the succession thickens to some 8 km, with little change of lithology. The subdivision and nomenclature of the Huronian has been modified in recent years and Table 4.2 shows the terms used by Roscoe (1968).

The lower part of the Mississagi Quartzite of Collins appears to be largely fluviatile and to have buried an irregular pre-Huronian land-surface. It carries disseminated uranium minerals in quartz-pebble conglomerates, among which is the valuable deposit of Blind River. At higher levels in the Bruce, the sandstones are usually feldspathic or impure and of variable thickness; they are associated with polymict conglomerates carrying boulders of granite and greenstone. Dolomitic limestones occur at two levels. The Cobalt division begins with the well-known and widely-developed *Gowganda formation* in which bouldery horizons alternate with greywackes and other impure, sometimes pebbly, sandstones through a thickness of over 800 metres. The thick sandstone groups which form the upper part of the Cobalt are well sorted, often current-bedded and of shallow-water origin. A glacial origin was suggested by Collins for the Gowganda boulder-beds, and this interpretation is strongly supported by the discovery of what appears to be a glacial pavement beneath the lowest boulder-bed in some localities. The association of those beds with turbidites has, however, led some geologists to infer that the tilloids themselves are slump -rocks or turbidites. If the Gowganda formation is correctly interpreted as a tillite, it provides evidence of glaciation earlier than about 2200 m.y.

The Huronian, with its abundance of clastic sediments, appears to include erosion-debris derived from the Kenoran basement. This impression is confirmed by the orientation of current-bedding in the Mississagi and Lorrain quartzites which indicates a rather constant flow of palaeocurrents towards the south-east or south. From studies not only of these rocks but also of current-bedded

Table 4.2. THE HURONIAN SUPERGROUP
(The succession of the type region, based on Roscoe, 1968)

Group			Formation	Lithological types
BRUCE DIVISION	COBALT		Bar River	quartzite (siltstone)
			Gordon Lake	siltstone
			Lorrain	quartzite, arkose
			Gowganda	argillite, conglomeratic greywacke (tillite?), arkose
	QUIRKE LAKE		Serpent	arkose, sub-greywacke
			Espanola	dolomite, limestone, siltstone, greywacke
			Bruce	conglomeratic greywacke
	HOUGH LAKE	Mississagi Quartzite	Mississagi	coarse sub-greywacke
			Pecors	argillite, siltstone
			Ramsay Lake	conglomeratic greywacke
	ELLIOT LAKE		McKim	sub-greywacke,
			Matinenda	argillite, arkose, uraniferous conglomerate
			Copper Cliff	acid volcanics
			Thessalon, Pater, Stobie	basic volcanics
			Livingston Creek	arkose

psammites in the Animikie and its equivalents of the Lake Superior region (p. 76), Pettijohn (1957) concluded that, over a considerable period of time, the regional palaeoslope was inclined from the central part of the stabilised Superior province towards the south and south-east — that is, towards the site of the southern branch of the Hudsonian system of mobile belts (Fig. 4.4).

3 The Animikie group

The southern side of the Superior province is flanked by a region (the 'Southern province' of Canadian terminology) strongly folded and metamorphosed in varying degrees during the Hudsonian cycle. The early Proterozoic Animikie Group, very broadly equivalent to the Huronian, forms a thick cover-succession within this mobile belt and overlaps locally onto the Archaean foreland as a thinner and less disturbed platform-sequence. The palaeoslope defined by current-bedding in the main psammites is inclined southward, indicating that detritus was transported from the craton towards the mobile belt (Fig. 4.4). Banded iron formations occur both in the platform-succession to the north of Lake Superior and in the succession of the mobile belt in Michigan and

Wisconsin, where they provide the iron ores which supply the industrial complexes of Detroit and the surrounding regions.

To the north of Lake Superior, a thin basal conglomerate is followed directly by the *Gunflint iron formation* which in turn is followed by carbonaceous pelites. Well preserved micro-organisms in the undisturbed cherts of the Gunflint have provided valuable material for the study of early life (p. 185).

To the south of Lake Superior, the succession is thicker and more complex. The lower formations are shallow-water orthoquartzites, dolomites and slates which are followed by a division of cherty iron formations, slates and basic volcanics and lastly by fine-grained greywackes, the sequence amounting in some regions to a total of 10 km. The iron formations are characterised, like other early Proterozoic iron formations, by the alternation of chert layers with layers in which the chert is associated with magnetite or with iron carbonates or silicates. They reach individual thicknesses of up to 300 m in the folded belt south of Lake Superior, where they form the Mesabi, Cuyuna, Penokee and Marquette ridges. The ore-bodies in these formations are enriched in hematite or goethite as a result of processes operating during or after the period of folding.

4 The Kaniapiskau Supergroup, Labrador trough

In the north-eastern branch of the Hudsonian system of mobile belts, a cover-succession – the Kaniapiskau Supergroup – occupies a tract some 100 km broad along the border of the Superior province. The outcrop of this succession runs south-south-east through Labrador to the Grenville front and then turns southward through the Grenville belt as a series of irregular and discontinuous outcrops.

The Kaniapiskau succession can be traced for considerable distances parallel to the length of the Labrador belt, but shows marked lateral variations which are considered to be related to the development of two parallel basins of deposition. These variations are illustrated in Fig. 4.5 based on a recent synthesis by Dimroth (1970). On the western side of the belt, a succession of miogeosynclinal type rests on a little disturbed Archaean basement and overlaps onto the foreland. The principal members of this succession are orthoquartzite–dolomite groups associated at two levels with banded iron-formations and shales which pass upward into pelites with the characters of flysch. In the eastern zone a thicker eugeosynclinal succession appears (Table 4.3), in which sedimentary divisions broadly similar to those of the west alternate with, and are followed by, tholeiitic pillow-lavas and pyroclastics. Thick sheets of gabbroic and ultramafic rocks intrude the volcanic groups and the underlying sediments, bringing the total thickness of igneous material in the basin-fill up to well over 10 km. The succession of the eugeosynclinal basin rests on a basement which is partially or entirely modified by Hudsonian reactivation, and is itself metamorphosed to varying degrees (Fig. 4.5). A zone of antiformal character separating it from the less disturbed miogeosynclinal zone to the west is regarded by Dimroth as marking the site of a geanticline developed during the period of deposition.

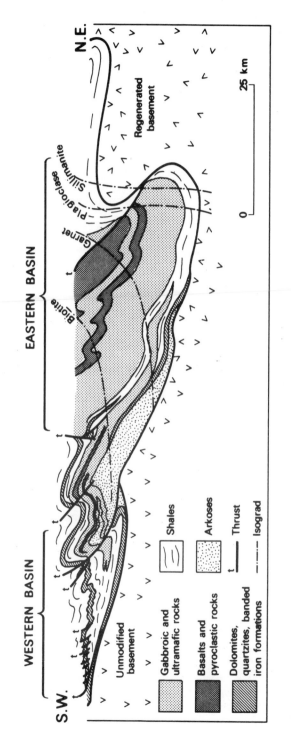

Fig. 4.5. Diagrammatic cross-section of the Labrador trough (based on Dimroth, 1970)

Table 4.3. THE KANIAPISKAU SUPERGROUP OF THE CENTRAL ZONE IN
THE WAKAUCH LAKE AREA (based on Baragar, 1967)

Doublet Group	*Willbob Formation*	mainly basaltic lavas (5000 m)
	Thompson Lake Formation	mainly slates, siltstones, quartzites, greywackes (500–700 m)
	Murdoch Formation	basic pyroclastics, minor basalts, acid tuffs, sediments (1000–2000 m)
Knob Lake Group	*Menihek Formation*	black shale, impure psammites, minor basalts (5000 m)
	Purdy Formation	dolomite (>500 m) *(disconformity)*
	Sokoman Formation	banded iron formation (150–250 m)
	Ruth and Wishart Formations	psammites and pelites with banded iron formations (>150 m)
	Fleming Formation	mainly cherts (>150 m)
	Denault Formation	dolomite with stromatolites, chert (0–1400 m)
	Attlikamagen Formation	pelites, psammites, basalts (350–800 m) underlain by dolomites and cherty psammites (?3000 m)
	Seward Formation	mainly red beds (600–1500 m)

The broad structure of the Labrador belt as interpreted by Dimroth and others bears a close resemblance to the classic model of a geosyncline as exemplified by the Appalachian belt. The miogeosynclinal zone flanking the foreland, occupied mainly by a sedimentary fill, is non-metamorphic and shows comparatively simple folding associated with thrusting. The eugeosynclinal zone, separated from it by the remnants of an early uplifted tract, contains vast thicknesses of basic igneous rocks of ophiolitic affinities which chemically resemble the oceanic tholeiites; the rocks of this zone are metamorphosed and are underlain by a regenerated basement.

As we have already seen, the Kaniapiskau Supergroup forms part of a series of supracrustal sequences which fringe the Archaean craton of the Superior province. Sequences bordering the smaller cratonic block of the Slave province are represented in the *Coronation geosyncline* which occupied much of the Bear province (Fig. 4.2). Hoffman (1973) has described a shelf sequence of quartzites and stromatolitic dolomites, passing westward into much thicker sequences of basic lavas followed by turbidites perhaps deposited near a continental margin. Rocks of the western type are folded, metamorphosed up to sillimanite grade and overridden by granitic massifs whereas those of the eastern type remain largely unaltered but are disrupted by numerous thrusts.

5 Hudsonian orogenic activity

The cover-successions of early Proterozoic age which lie within the Hudsonian mobile belt system are conspicuously folded, providing a contrast to the equivalent successions which rest undisturbed on the Archaean of the foreland regions. The presence of thick competent units such as the iron formations led to the development of folds of large wavelength some of which can be traced for long distances. Such large folds, dissected by erosion, are well seen in the 'iron ranges' south of Lake Superior. The term *Penokean orogeny* used especially in the United States for the folding and metamorphism in this region is, very broadly, equivalent to the term Hudsonian orogeny.

The metamorphic grade of the cover-rocks is low in the marginal parts of the mobile belts and rather variable within them. Over parts of the Labrador trough the grade rises to granulite facies, but declines both eastward and westward towards the margins of the mobile belt. To the south of Lake Superior and, at least locally, north-east of Lake Huron, the characteristic mineral assemblages are those of a low-pressure facies series. Pelitic rocks of the Animikie Group contain andalusite and' staurolite in the middle grades and sillimanite and staurolite at high grades. The zones of high-grade metamorphism in the Lake Superior region form 'nodes' often no more than 25 km in diameter which are independent of the fold-structures and are characterised by an abundance of granitic material. The narrowness of the metamorphic zones, the occurrence of low-pressure assemblages and the abundance of granite – all features which recall the characters of the Svecofennide belt – suggest that geothermal gradients in the mobile belt were high.

The large areas affected by Hudsonian plutonism, especially those in the western part of the Churchill province and the eastern Nain province, which are made principally of granites, gneisses and highly altered supracrustals, are less easily interpreted.They undoubtedly contain numerous granite bodies which were emplaced during the Hudsonian orogeny and which correspond to the Hudsonian or Penokean granites intruded into the Kaniapiskau and Animikie groups. In some regions, traces of Archaean tectonic patterns can still be detected and certain rocks have yielded radiometric dates substantially older than the range 1900–1750 m.y. characteristic of the Churchill province. Such features are seen locally at the borders of the Slave and eastern Nain province.

In the interior of the western Churchill province considerable areas are occupied by granites and gneisses traversed by greenstone belts not unlike those of the adjacent Archaean terrain, though generally without their characteristic gold mineralisation. The aeromagnetic map of the western Churchill province displays a pattern of east—west anomalies similar in scale and intensity to those of the Superior province, traversed by and deflected by linear tracts of anomalies with consistent north-easterly trend (Fig. 4.6). This large-scale superposition of structures suggests that modified Archaean rocks are widespread in the deeper parts of the Churchill province and that tectonic reworking of these rocks involved the development of large, probably transcurrent, dislocations. The *Nelson River* lineament, the marginal feature of the province south-west of Hudson Bay, is the site of one such linear disturbance and is marked by obvious complexities of geological structure. A belt of granulite-facies gneisses and anorthosites approaching this lineament from the east is deflected southward and retrogressed. It terminates close to a shear-zone showing a strong north-easterly foliation which dips south-eastward towards the Archaean foreland region. A steep gravity gradient from a high bouguer anomaly over the rocks of granulite facies (the 'Nelson River high') to low anomalies related to granitic bodies near the dislocation zone, marks the main boundary between

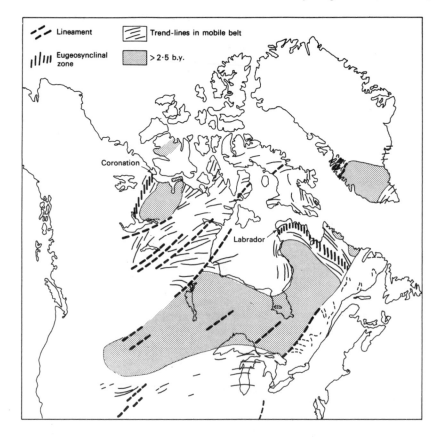

Fig. 4.6. Some major lineaments in the Precambrian of North America; indications of transcurrent displacements characterise some of the structures shown (based on Sutton and Watson, 1974)

provinces characterised by Hudsonian and Kenoran radiometric ages. Near the border zone an important string of nickeliferous peridotites is emplaced. The location of these bodies along the discontinuity, indicated both by the structures on the ground and by the gravity and aeromagnetic data, confirms the tectonic importance of the front of Hudsonian regeneration.

6 Mineralisation of the Hudsonian cycle

The main economic importance of the early Proterozoic provinces in North America depends on the banded iron formations which have already been discussed. These chemical sediments are located in the mobile belts or along the forelands adjacent to them. Uranium is important in the Bear province where many small pitchblende occurrences have been recorded and in Huronian conglomerates resting on Archaean rocks of the Superior province (p. 74). Other metals occurring mainly within outcrops of early Proterozoic supracrustal rocks are copper, lead, zinc and native silver (the latter located in the Nipissing diabase sills).

7 Late-orogenic events

The termination of Hudsonian orogeny in the western part of the Churchill province was rapidly followed by the accumulation of sediments of a facies, which suggests that mountain-building in the literal sense had preceded stabilisation. The *Dubawnt Group*, which rests unconformably on crystalline rocks of the province in the North-west Territories, is made predominantly of clastic sediments and volcanics. Its lower division, laid down only locally, consists of immature red-beds characterised by polymict conglomerates regarded by Donaldson as piedmont deposits. A more extensive middle division consisting of intermediate and alkaline volcanics is followed by an upper division of better-sorted sandstones. An isochron age of 1735 m.y. for some of the volcanics shows that the group began to accumulate very soon after the termination of orogenic activity and justifies a broad analogy with post-tectonic molasse-formations of younger orogenic belts.

8 Anorthosites and their associates

The eastern part of the Canadian shield contains a number of large rounded or ovoid masses of anorthosite, with maximum diameters of up to 200 km, which are emplaced in a zone extending from the Labrador coast through the Grenville province to the Precambrian inlier of the Adirondack mountains. The principal mineral of these rocks is labradorite or basic andesine; plutonic rocks associated with them include gabbroic, syenitic and granitic types (including rapakivi) and are frequently characterised by orthopyroxene.

Because the majority of the Canadian anorthosites fall within the Grenville belt, they have sometimes been regarded as associates of this belt. Recent studies of the Adirondacks anorthosites and of some other masses in the Grenville province, however, suggest that these are reactivated bodies which were partially mobilised during the Grenville cycle to form structures analogous to mantled gneiss domes (e.g. Walton and de Waard, 1963). Anorthosites in Labrador which lie outside the Grenville province retain primary structures and appear to have been intruded after the peak of Hudsonian orogenic activity: one of these unmodified bodies has yielded an age of 1400 m.y. Both in timing and in geological relationships, the eastern Canadian anorthosites and their associates have features in common with the late Ketilidian rapakivi-anorthosite suite of Greenland; both may belong to an igneous province originally extending eastward into the parts of Fennoscandia characterised by rapakivi granites, (p. 35).

VI The mid-Proterozoic Craton

With the termination of Hudsonian orogenic activity at about 1750 m.y. and the subsequent rather protracted stabilisation of the Hudsonian provinces — in those affected by the so-called Elsonian cycle extending down to about 1400 m.y. — a very large crustal region passed into a condition of stability. This phase of mid-Proterozoic stabilisation led to the establishment of a tectonic regime foreshadowing that which ruled in North America throughout Phanerozoic times: a large central craton came into existence, flanked on the south-east by the Grenville mobile belt and on the south-west by a zone of subsidence from which, ultimately, the mobile belt of the Rocky Mountain system emerged.

1 Cratonic cover-successions

Before turning to the Grenville belt, we may deal briefly with some features of the central craton. Later Proterozoic cratonic cover-successions are extensively developed in the south but are rather restricted in the north — a distinction which suggests that the shield-region of Canada was already differentiated from the platform-region of the United States (Fig. 4.1). We may mention some representative formations in each region, starting in the north.

The *Athabasca formation* to the south of Lake Athabasca rests unconformably on crystalline rocks dated as Hudsonian and is itself cut by basic dykes of the Mackenzie swarm (p. 83). The bulk of the formation consists of clastic sediments which are predominantly well sorted quartz-sandstones with lenses of oligomict conglomerate; but dolomitic limestones appear above the sandstones and there are also intercalations of basic volcanics. The primary structures — festoon and planar cross-bedding and ripple-marking in the sandstones and oolitic and stromatolitic structures in the dolomites — agree in suggesting shallow-water deposition. Fahrig (1961) concludes that the formation may have been laid down partly on a coastal plain and partly in shallow sea. The sediments are of orthoquartzite facies, well graded and well sorted; but immediately to the north of the Athabasca formation lie outcrops of immature piedmont deposits — the *Martin formation* — which, though probably somewhat older, belong to the same general period of deposition. Remarkably constant palaeocurrent directions determined from current-bedding throughout the Athabasca formation indicate a palaeoslope towards the west or north-west.

In the vicinity of Lake Superior and in a tract extending southward beneath the Phanerozoic cover for nearly a thousand kilometres, the late Proterozoic succession is dominated by plateau-basalts. This succession forms the *Keweenawan* which rests unconformably on the crystalline basement and the Animikie formations. Its lower and middle divisions are older than about 1100 m.y., the age of the Duluth gabbro which intrudes them. The maximum thickness of the Keweenawan is of the order of 15 km, of which at least half is of volcanic origin. The volcanic rocks are plateau-basalts, mainly in the form of thin lava-flows, often amygdaloidal and carrying native copper and copper sulphides. The sedimentary components, which are interbedded with the lavas in the lower divisions and form most of the upper divisions, are reddish sandstones, conglomerates and mudstones showing shallow-water depositional features and are thought to represent flood-plain sediments.

The Keweenawan volcanic outcrops of Lake Superior, themselves of considerable bulk, are only a portion of a much more extensive province of plateau-basalts which occupies a narrow tract (possibly a graben) extending beneath the post-Cambrian cover as far south as Kansas. This tract, which has been identified in a number of boreholes, is defined by strong magnetic and gravity anomalies (the 'mid-continent gravity high'). Similar basaltic piles – also characterised by copper mineralisation – occur in the Coppermine River area of the North-west Territories of Arctic Canada and major plutonic basic complexes and regional dyke swarms of much the same age are represented in the Canadian shield. Taken together, these occurrences record a major period of basic magmatism of cratonic type at about 1300–1050 m.y.

The later Proterozoic cratonic rocks of the central and southern United States, unlike those of the Canadian shield, are sufficiently extensive to blanket the crystalline basement over large areas. Two principal components seem to bulk large in this cover – an assemblage of acid volcanics and a spread of shallow-water clastic sediments. The *acid volcanics* which include rhyolites, pyroclastics and near-surface granitic bodies, are widespread in the southern States, where they may be post-orogenic volcanics of the Hudsonian-Elsonian cycle. The clastic sediments are represented on the craton, for example, by the *Sioux Quartzite* of Minnesota and neighbouring states, an orthoquartzite giving a minimum age of about 1200 m.y. More important in terms of thickness and extent are the later Proterozoic deposits of Montana and other parts of the western mobile belt; the Precambrian *Belt Series* and *Purcell Series*, span very long periods (their oldest divisions dating back to 1300 m.y.) and reach total thicknesses of up to 12 km.

2 Basic intrusives of the craton

Two large bodies of differentiated gabbroic rocks are associated with the basic volcanics just referred to. The *Duluth gabbro*, near the western end of Lake Superior, is a stratiform body made up of ultrabasic cumulates, gabbros and acid differentiates. It is dated at 1100–1050 m.y. The *Muskox intrusion* of the North-west Territories forms a linear outcrop resembling an enormous roofed dyke up to 5 km in width. The gabbros, peridotites and picrites of the complex often show a vertical banding attributed to magmatic flow; but in the broadest part of the intrusion – which widens abruptly upwards – a horizontal layering produced by gravity-settling is developed, recalling the Great Dyke of Rhodesia (p. 113).

Even more spectacular are the *regional dyke swarms* of dolerites and basalts which appear to have been emplaced within the period 1300–1000 m.y. The Mackenzie swarm and its associates crosses the Canadian shield from north-west to south-east, extending through a tract some 2500 kilometres long and 400 km broad (Fig. 4.7). We may recall the swarm of similar age (the Gardar dykes) in southern Greenland.

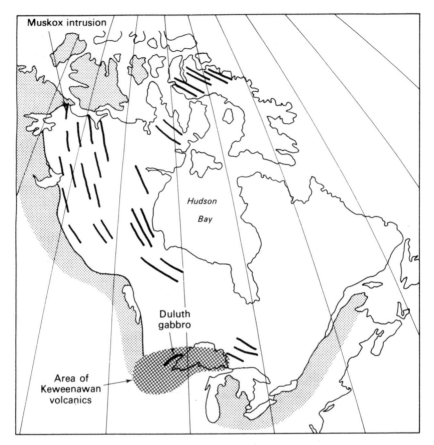

Fig. 4.7. Mid-Proterozoic cratonic igneous rocks in northern North America; the bold lines represent basic dykes (including the Mackenzie swarm) emplaced within the interval 1300–1000 m.y.

Finally, this seems the least inappropriate place to mention the *Sudbury igneous complex*, a basic plutonic body emplaced in Huronian sediments near the Grenville front, which is a major source of nickel. The complex crops out as an elliptical ring surrounded by Huronian rocks and enclosing outcrops of post-Huronian volcanics and sediments, the oldest of which have been dated at 1720 m.y. The outer (and probably lower) zones of the Sudbury complex are made of basic hypersthene-bearing rocks (*norites*), the inner of granophyric rocks which are capped by consanguineous rhyodacitic tuffs. Nickel-bearing sulphide deposits are emplaced near the outer margin of the complex. Hypotheses as to the origin of the Sudbury intrusive are many. Its elliptical form and the synformal arrangement of the supracrustal rocks at its centre led early workers to an interpretation in terms of the differentiation of a lopolithic sheet. More recently Williams outlined an interpretation in terms of cauldron subsidence, regarding the inward dips as those of the walls of a ring-complex. Finally, Dietz has proposed that the complex represents the result of meteor impact, leading to fusion of material in depth and emplacement at higher levels.

VII The Grenville Belt

The clearly defined Grenville province in the east of the Canadian shield (Fig. 4.2) was recognised as a structural entity at an early date and was for some time regarded as one of the oldest portions of the shield. More recent work has shown that although the province incorporates much regenerated Archaean and early Proterozoic rock it was not finally stabilised until a comparatively late stage. Isotopic dates for rocks and minerals of the province cluster about the period 1100–900 m.y.

The Grenville belt is bounded on the west by a well-defined orogenic front. It continues south-westward for a total distance of at least 3500 km to the Mexican border, but its southern parts are largely obscured by Phanerozoic rocks. It is flanked by, and partially obliterated by, the younger Appalachian mobile belt which follows an almost parallel course from Newfoundland to the Gulf of Mexico.

The high grade of metamorphism and migmatisation which is characteristic of the Grenville belt has to a large extent obscured its early history. There seems to be no general agreement as to the extent to which supracrustal rocks deposited during the early stage of the Grenville cycle are represented. Many metasedimentary relics have been recognised but, as will be shown immediately, some appear to correspond to pre-Hudsonian groups and should therefore be assigned to the basement rather than the cover. It is nevertheless held by some authorities that a *Grenville Series* directly connected with the Grenville belt is incorporated in the metamorphic and migmatitic complexes. In southern Quebec and adjacent parts of Ontario, near the type locality of Grenville, the assemblage of supracrustal rocks which form the host-rocks of migmatites includes conspicuous limestones, pure quartzites, pelitic gneisses and sedimentary amphibolites. It will be noted that this assemblage is rich in resisters to migmatisation. In the neighbouring Hastings region, where the metamorphic grade is lower and migmatisation less extensive, limestones and quartzites are associated with considerable thicknesses of pelites, arkoses and conglomerates. The original relationships of the metasediments in both areas are doubtful.

In the Precambrian inlier of the Adirondacks, marbles, amphibolites and other metasediments commonly referred to as the Grenville Series are associated with the Adirondacks anorthosite and related gneisses (Fig. 4.8). They have been regarded by Buddington, Balk and others as the country rocks into which the anorthosite was intruded; but, on the other hand, Walton and de Waard (1963) have argued that they were deposited on and subsequently interfolded with a basement of anorthosite and gneiss. From the dates tentatively assigned to the anorthosites (p. 181) it will be apparent that this latter hypothesis would imply that the Grenville Series of the Adirondacks is younger than about 1400 m.y., and therefore post-Hudsonian.

In other regions, formations deposited prior to the Hudsonian orogeny have been traced from the foreland into the marginal parts of the Grenville belt and it has been suggested that these formations are equivalent to the Grenville Series. In a classic memoir published in 1930 and entitled 'The disappearance of the Huronian', Quirke and Collins argued that members of the Huronian succession to the north-east of Lake Huron could be matched with metasedimentary relics enclosed in granites and gneisses lying some 5–10 km within the Grenville front. More recently, members of the Kaniapiskau succession of the Labrador trough have been traced for many tens of kilometres within the Grenville belt (Gastil *et al*, 1960); the iron formations in this succession provide easily identified key-horizons which make correlation across the Grenville front reasonably

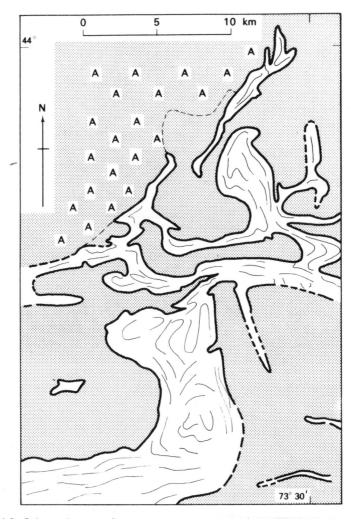

Fig. 4.8. Schematic map of the eastern Adirondacks (after Walton and de Waard, 1963) showing the supposed basement in relation to metasediments of the Grenville Series (A = anorthosite; other basement rocks are tinted)

secure. Both the Huronian and the Kaniapiskau contain carbonate rocks and quartzites and the suggestions have been made (e.g. Stockwell, 1964), first, that the outcrops of the two formations were originally linked by an arcuate zone of sediments extending through the Grenville belt and, second, that the surviving relics of these sediments constitute the Grenville Series. This interpretation would involve assigning most metasediments in the Grenville belt to formations which had been deposited during earlier geological cycles and would leave little evidence to suggest that a geosynclinal basin of deposition was ever developed on the site of the Grenville mobile belt.

The structural pattern of the Grenville belt as revealed both by surface mapping and by regional aeromagnetic surveys somewhat resembles that of the Svecofennides in that it is characterised by complex interference-structures and by granitic, migmatitic and anorthositic domes. The reactivation of old plutonic bodies in the basement, with the production of mantled gneiss domes, has been invoked to explain the relationship of the Adirondacks anorthosite with the Grenville Series and, since basement rocks appear to have been reworked on a regional scale, similar phenomena may be widespread. Spectacular interference-patterns have been demonstrated where the iron formations of the Labrador trough enter the northern part of the Grenville belt.

The *Grenville orogenic front* which forms the north-western boundary of the Grenville province was recognised as a fundamental line separating regions of contrasting structure and lithology at an early date in the geological investigation of the Canadian shield. In some localities, this front is marked by definite dislocations; elsewhere, it may be best interpreted as essentially a front of metamorphism or migmatisation at which rocks pass through a narrow transition zone from the condition characteristic of the foreland to that characteristic of the Grenville belt. Large folds and dislocations of Grenville age, sometimes associated with considerable disruption, may affect the rocks for some kilometres to the north-west of the front; the development of minor structures of plastic style which define the fabric of the Grenville belt is correlated with the rapid increase in metamorphic grade and the amount of migmatisation at the Grenville front itself. The close association between metamorphic grade and structural style limited the complete tectonic reworking of pre-existing rocks to the areas which reached high temperatures during the period of deformation; the structural front is therefore in effect an isograd as well as a zone of dislocation.

Along the entire length of the Grenville belt exposed in Canada, and in outcrops of Grenville rocks further south within the Appalachian belt, gneisses and migmatites are widely developed and metasedimentary rocks seldom retain primary structures. Metamorphism is of granulite facies over considerable areas in southern Canada and the Adirondacks; elsewhere it is generally of amphibolite facies. Granitic material appears both in regional migmatitic complexes and also in discrete intrusive bodies. The Grenville front is followed for over 40 km north-eastward from Killarney Bay on Lake Huron by a strip of granite seldom more than a kilometre or so wide. Both the granites and the metamorphic complexes are essentially barren so far as metallic ore-deposits are concerned.

An unusual feature of the Grenville belt is the development of syn-orogenic *nepheline-syenites* which are in part of metasomatic origin. The nepheline-syenites of the Bancroft area in Ontario, described in a classic memoir by Adams

5

Precambrian of Asiatic Laurasia

I Preliminary: The Make-up of Asia

The geological constitution of so vast a land-mass as Asia can only be dealt with here in the broadest outline, to provide a background for the discussion of topics of special interest which are considered in later pages. The continent of Asia as it now exists is composite; part is derived from Laurasia, part from Gondwanaland, the two portions being separated by the broad Himalayan mobile belt (Fig. 5.1).

To the south of the mobile belt lie the two ancient stable blocks of *Peninsular India* and *Arabia* which are regarded as fragments of Gondwanaland. The geological history of these blocks through Precambrian, Palaeozoic and Mesozoic times had more in common with the history of Africa than with that of the remainder of Asia. It appears that they were not united with the remainder of Asia until late in the Phanerozoic.

The complex *Himalayan mobile belt* enters Asia from the Mediterranean, continues through the Middle East to the Himalayas, and then turns sharply southward through Burma, Malaysia and the island arcs of Indonesia. It is still a zone of considerable orogenic activity and appears to have been active intermittently throughout the Phanerozoic.

To the north and east of the mobile belt lies the main mass of *central and northern Asia* which is regarded as a part of the Laurasian continental group. This vast and complex region at present constitutes a stable block, the largest in the world, but it is intersected by stabilised orogenic belts in which activity continued until midway through the Phanerozoic. We may therefore regard it as a patchwork of ancient cratons and early Phanerozoic mobile belts. In the north the *Siberian platform* has a Precambrian basement which emerges at the surface in several massifs, but is largely covered by great thicknesses of more recent sediments and volcanics. Much of the platform has suffered only limited warping since the beginning of Phanerozoic time, but its south-western border is

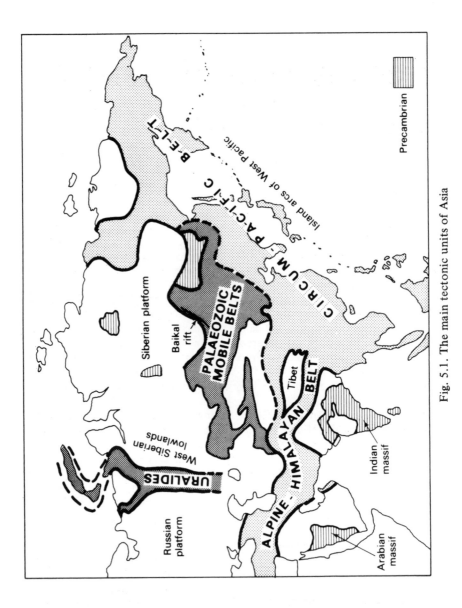

Fig. 5.1. The main tectonic units of Asia

traversed by the great *rift-valley system of Lake Baikal*. To the west of the platform a colossal downwarp beneath the *West Siberian Lowlands* contains a thick but little disturbed assemblage of Mesozoic and Tertiary sediments. This in turn is limited on the west by the Upper Palaeozoic fold-belt of the *Ural Mountains*. To the south of the Siberian platform is a complex region of elevated mountain-tracts and depressed basins composed of rocks which though folded mainly in Palaeozoic times were also subjected to more recent vertical movements. To the east of the platform lies a complex of largely Mesozoic fold-belts constituting much of north and east China. Finally, at the eastern margin of the Asian continent are the fringing island arcs of the still-active *circum-Pacific mobile belt*.

The Precambrian regions of Asiatic Laurasia are small by comparison with the great Precambrian shields and are separated from these shields by thousands of kilometres of younger rocks. Accordingly, we shall deal with these regions rather briefly and for the sake of convenience we shall include in this chapter not only accounts of the Precambrian massifs which have remained stable throughout Phanerozoic times but also some reference to the Precambrian rocks incorporated in younger fold-belts. The Precambrian shield of Peninsular India must logically be considered as a fragment of Gondwanaland and is therefore dealt with in a later chapter.

II The Precambrian of Asiatic Laurasia

The Precambrian areas to be mentioned can be divided on a regional basis into:

(a) those appearing in and on the borders of the *Siberian platform*; and

(b) those of *East China, Manchuria and Korea* which lie within or between the marginal Phanerozoic fold-belts.

The time-division of Precambrian rocks employed in the U.S.S.R. is based on the recognition of an earlier *Archaeozoic* and a later *Proterozoic* assemblage, each being grouped in many regions into two, or sometimes three, divisions. The uppermost part of the Upper Proterozoic is commonly closely associated with undoubted Cambrian. Rocks in this stratigraphical position have been encountered in deep borings beneath the Russian platform, in the Siberian shield and in China where they include horizons interpreted as being of glacial origin. This division, falling near the boundary between Precambrian and Cambrian, has analogues in many other parts of the world. Its nomemclature is somewhat confused. On the splendid geological map accompanying Nalivkin's (1960) *Geology of Russia* it is designated $Pr_2 Cm_1$, indicating Upper Proterozoic or Lower Cambrian. After the map had been prepared, it was given the name *Sinian complex* and in the legend of the map was placed between the Proterozoic and Cambrian as a separate division. More recent publications (e.g. Keller, 1964, Sokolov, 1964) use the term *Vendian complex* to cover what appear to be equivalent groups in the Urals, Kazakhstan and Siberia, giving a lower age-limit of 600–650 m.y., which is close to the date generally given for the base of the Cambrian. It is worth noting the occurrence of tillites in units of this general age in the Tien-Shan range and elsewhere.

Another term widely used in the U.S.S.R. is *Riphean* which, again, appears to have been employed in several different senses. Some authors equate Riphean with the Sinian complex. More recent papers, such as that of Keller (1964), include in the Riphean, groups deposited over the period 1550–650 m.y. and thus use the term to include most Proterozoic supracrustal formations. Keller equates Riphean, in this sense, with the Sinian as originally instituted by Grabau. A threefold subdivision into Lower, Middle and Upper Riphean is commonly adopted.

In the light of the above, we arrive at a rough classification of the Precambrian rocks of Asiatic regions, which is given with some supplementary details in Table 5.1.

Table 5.1. PRECAMBRIAN DIVISIONS OF ASIATIC LAURASIA

Dates m.y.		Stratigraphical terms		Orogenic episodes
600		Vendian = Sinian		
850	PROTEROZOIC			Baikalian
1100			Upper	
1350		Riphean	Middle	Ovruch, Volynian
1550			Lower	
c. 2000				Svecofennide
	ARCHAEO–ZOIC			Belomoride
				Saamide

III Precambrian Massifs of the Siberian Platform

The great Siberian platform is underlain by Precambrian rocks that had completed their orogenic history by the close of Proterozoic times. Some Russian geologists regard the platform, like the Russian platform to the west of the Urals, as a Precambrian geosyncline. Other authorities consider the central portion to be a *Siberian shield* or nucleus, against which successive fold-belts consolidated to produce the continental mass of northern Asia.

The main exposed areas of Precambrian in and at the margin of the Siberian platform are, first, the predominantly Archaeozoic massifs of Anabar and Aldan and, second, the Proterozoic of the south-western frame of the platform (Fig. 5.2).

The *Anabar massif* forms an elevated region around the headwaters of the Anabar River in the north of the platform. Archaeozoic crystalline rocks appearing over an area of some $60\,000\ km^2$ are seen framed by an uncon-

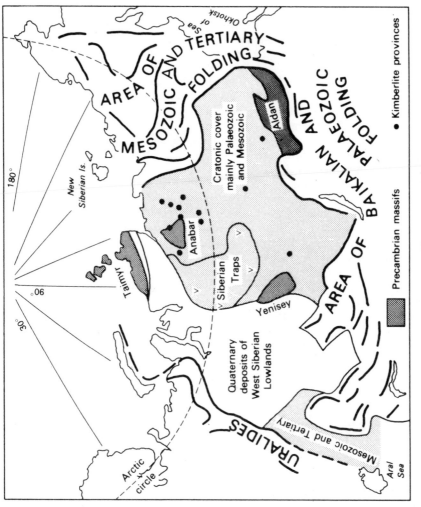

Fig. 5.2. The Siberian craton with the Phanerozoic mobile belts at its margin

formable and flat-lying cover of Sinian sandstones, shales and limestones which passes up conformably into the base of the Cambrian. Here, clearly, stability was attained before the end of Precambrian time.

The Archaeozoic of the massif is composed principally of high-grade gneisses, schists and migmatites folded on north-westerly axes and traversed by crush-zones with a somewhat similar trend. The metamorphic rocks are predominantly of granulite facies, characterised by the wide distribution of hypersthene and the local abundance of garnet and scapolite. Metasedimentary relics include marbles and calc-gneisses, with quartzites and pelites carrying graphite, sillimanite or cordierite. Granites and granodiorites form large elongated masses parallel to the tectonic grain. Smaller basic and ultrabasic intrusions are also recorded.

The *Aldan massif* occupies an area extending eastward for some 1000 km from the River Vitim to the Sea of Okhotsk in eastern Asia. It carries outliers of unfolded Sinian and passes northward beneath the Cambrian and younger cover of the Siberian platform. It is truncated on the west, south and east by fold-zones of Phanerozoic age. The southern margin is a dislocation-zone along which the Precambrian of the massif is brought in many places against Proterozoic modified by Palaeozoic movements. Within the massif, the predominant strike is north-north-west.

The oldest rocks of the massif, regarded as Lower Archaeozoic and yielding U–Pb ages of around 3000 m.y., include high-grade gneisses, migmatites and charnockites. Younger divisions, also largely of granulite facies, are seen in the valley of the Aldan River where metasediments including thick crystalline limestones, quartzites and pelites are recorded. Sillimanite, sometimes associated with cordierite, is a common mineral of the pelites. Ultrabasic bodies intrude the gneiss complexes. Granites include both concordant types interpreted as of anatectic origin which often form dome-like bodies, or occupy the cores of migmatite zones and discordant intrusions. Dating of minerals from both supracrustal rocks and the associated granites has yielded minimum ages of about 2000 m.y., indicating the approximate time of stabilisation. The gold deposits of the Aldan and adjacent areas, which include both lodes and placers, are situated in or derived from these ancient rocks.

On the south-eastern and south-western borders of the Siberian platform, enormous areas of more or less modified *Proterozoic rocks* with smaller developments of Archaeozoic appear within the belt of complex tectonics which frames the platform (Fig. 5.2). From east to west, the main Precambrian regions include the mountain-tract south of the Aldan massif, the arcuate belt extending from north-east of Lake Baikal along the lake into the Sayan mountains, some massifs further north along the Yenisey River, several blocks in Kazakstan and others in the southern Urals. In many of these regions there is evidence of geological cycles dating from about 1600 m.y. down to 800 m.y., suggesting that one or more Proterozoic orogenic belts are represented. The last Precambrian cycle, the *Baikalian*, appears to have come to an end at about 800 m.y. The dominant grain of the structures roughly follows the course of this belt, but this may be at least in part a result of Palaeozoic activity.

Proterozoic metasediments and metavolcanics, some sufficiently well preserved to be assigned to appropriate divisions of the Riphean, are represented in many regions. An example of a well-dated succession is given in Fig. 5.3. Locally, as to the south of the Aldan massif, Archaeozoic rocks showing evidence of retrogressive metamorphism have been reported and dates of more than 3000 m.y. have been obtained. In general, the grade of metamorphism appears to be high and rocks of granulite facies occur in the Baikal region. Anorthosites dated at 2300 m.y. occur in the area south of the Aldan massif. Granitic rocks of various types appear to be extremely widely-developed, forming vast bodies often elongated parallel to the tectonic trend.

Far to the north, a Precambrian massif separated from the Siberian platform only by an extension of the Siberian lowlands, makes part of the *Taimyr Peninsula*. This massif is incorporated in a Palaeozoic fold-belt regarded as an

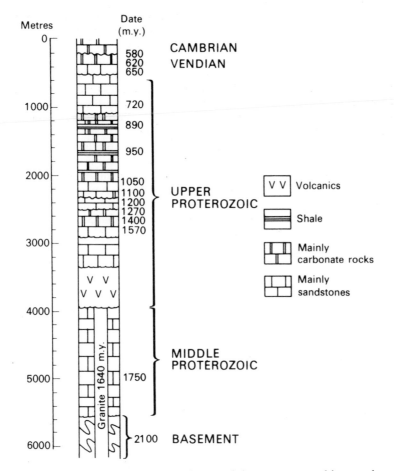

Fig. 5.3. A dated Proterozoic succession overlying a metamorphic complex near the southern border of the Aldan shield (after Kazakov and Knorre, 1970). The column gives a composite sequence pieced together from several localities and shows the positions of isotopically dated key-beds (dates are mainly K–Ar determinations of glauconites)

extension via Novaya Zemlya of the Urals. Its rocks are extensively sliced up by dislocations of ENE trend, some of which carry Precambrian wedges southward towards the stable region. The metamorphic complex of Taimyr includes rocks showing a great range of metamorphic states, from gneisses to schists and phyllites.

IV The Precambrian of Eastern Asia

In eastern China, Manchuria and Korea, the Phanerozoic fold-belts of eastern Asia incorporate a number of relatively stable blocks in which recognisable Precambrian rocks are preserved. At least two Precambrian orogenic cycles have been recognised, for which a number of local names have been employed. In Shan-si, for example, an older *Watai-Taishan* cycle terminating by 2500 m.y. affects a great variety of supracrustal rocks. Among those of sedimentary origin are mica-schists and gneisses, followed by quartzites, amphibolites of igneous origin and finally by slates, grits and limestones. A stromatolite, *Gymnosolen*, is widely distributed in the highest sub-division. A younger group includes biotitic and hornblende gneisses, pelitic schists and crystalline limestones. All these rocks were folded, metamorphosed in varying degrees, intruded by granites and subsequently uplifted and eroded.

The younger *Hutoian* cycle began with the deposition, on the peneplained rocks of the older cycle, of pelites, sandstones and banded ironstones followed by a series of prominent limestones and dolomites. Folding and granitic activity ended the cycle which possibly began at about 1100 m.y. A pegmatite attributed to the same cycle in Manchuria has given an isotopic age of 800 m.y. and a minimum stratigraphical age is also given by the fact that Sinian deposits rest unconformably on folded Hutoian.

6

The African Cratons

I Preliminary: The Make-up of Africa

The African continent (Fig. 6.1) has as its basement a huge platform of crystalline rocks partially masked by flat lying cover-successions of many ages. Towards the eastern, northern and western coasts, the crystalline complexes descend in many places beneath successions of Mesozoic and Tertiary sediments deposited in *marginal marine basins*. In the extreme north-west, a portion of the Alpide fold-belt makes the *Atlas Mountains*, while in the extreme south the *Cape fold-belt* represents a Palaeozoic mobile belt. The great fractures which define the *African rift-valley system* traverse the continent from north to south and their course is marked by *rift-volcanics* of Tertiary and Recent date.

The surface of Africa stands high — much of it at more than 1000 m — and, as will be seen in Part II, the continent has been subjected to uplift through the greater part of Phanerozoic time; it is, in fact, the largest persistent land-area of which we have detailed knowledge.

The crystalline basement was until recently thought to be made up almost wholly of Precambrian rocks. Radiometric dating has shown, however, that, in addition to large massifs of older material, it includes a network of belts which were not stabilised until early Palaeozoic times (Fig. 6.2). Prominent among these are the *Mozambique belt* which runs north and south near the eastern margin of the continent and the *Katanga belt* which crosses central Africa from east to west. In view of the late date of their final consolidation, we shall postpone consideration of the belts in this network until Part II of this book.

II The Older Precambrian Cratons

Three great massifs of older Precambrian rocks remained stable throughout the development of the late Precambrian—early Palaeozoic system of fold-belts just referred to. Each is roughly equidimensional and is of the same order of size as the Baltic shield. Rather extensive cover-sequences of late Precambrian or Phanerozoic age overlie parts of the massifs which, therefore, do not conform to

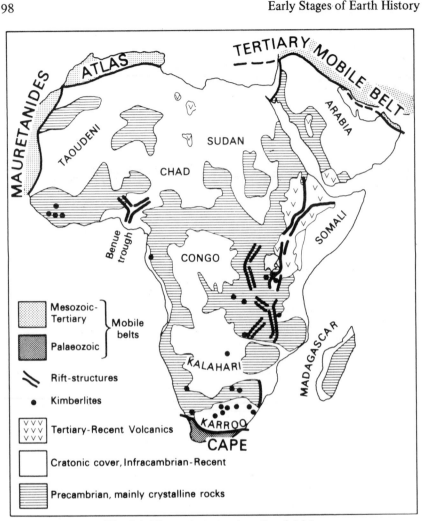

Fig. 6.1. The main tectonic units of Africa

our definition of shields; for this reason we shall refer to them as *Precambrian cratons.*

Following Kennedy (1965), we may use the names *Kalahari craton, Congo craton* and *West African craton* for the three major units outlined in Fig. 6.2: the principal provinces in each of these massifs are listed in Table 6.1. The Congo and Kalahari cratons are separated only by the comparatively narrow Katanga belt and will be dealt with together, leaving the more isolated West African craton to be described separately. As Table 6.1 shows, it is possible to draw a general distinction between *Archaean provinces*, which were largely stabilised by about 2600 m.y., and *early and middle Proterozoic belts* which were stabilised at various times between 1800 m.y. and 1000 m.y. We shall consider these two units in turn.

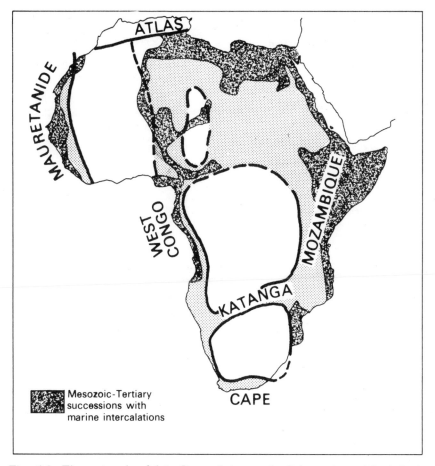

Fig. 6.2. The network of late Precambrian—early Palaeozoic mobile belts in Africa with the principal outcrops of marine Mesozoic and Tertiary sediments superimposed (based on Kennedy, 1965)

III The Archaean Provinces of Southern and Eastern Africa

The principal Archaean regions of the Kalahari and Congo cratons are, from south to north, the massifs of the *Transvaal, Rhodesia* and *Tanzania*. Smaller remnants of rocks yielding isotopic ages of 3500—2500 m.y. occur in the central region of Madagascar and in the West Nile area. The three large Archaean provinces have much in common. Each has a tectonic grain running roughly east and west and is made up largely of granitic or migmatitic rocks. Each incorporates irregular greenstone belts made up of slightly-metamorphosed supracrustal rocks which locally carry gold deposits. In many respects they resemble the Superior province of Canada and the Yilgarn and Pilbara blocks of western Australia. The Transvaal province is of particular interest for several

Table 6.1. THE MAIN PRECAMBRIAN PROVINCES OF THE AFRICAN CRATONS

A	*The Kalahari Craton*	Limpopo belt 2600–2000 m.y. Rhodesian province 3400–2600 m.y. Transvaal province 3400–2800 m.y. Orange River belt > 1000 m.y.
B	*The Congo Craton*	Karagwe-Ankolean belt 1400–1000 m.y. Toro belt 2000–1700 m.y. Ubendian belt 1800 m.y. Tanzania province 3000–2000 m.y.
C	*The West African Craton*	Birrimian belt 2200 m.y. Older complexes *c.* 2900 m.y.
D	*The late Precambrian–early Palaeozoic orogenic belts* enclosing the cratons and dated at *c.* 650–450 m.y.:	the Katanga belt, Mozambique belt, Zambesi belt and others (see Part II).

reasons: at least some of its supracrustal groups are extremely old, dating back to nearly 3400 m.y. and are locally almost unaltered, providing exceptionally favourable material for study; finally, parts of the massif received a unique succession of Archaean cratonic supracrustals. With these features in mind, we shall deal first with the crystalline province of the Transvaal, then with its Archaean cover and finally with the provinces of Rhodesia and Tanzania.

IV The Oldest Rocks of the Transvaal Province

In the eastern Transvaal and Swaziland, the Archaean crystalline basement represents a typical granite-greenstone belt terrain, bordered on the north by a highly-metamorphosed tract known as the *Limpopo belt* which will be discussed later (p. 112). In the main part, granites, gneisses and migmatites carrying remnants of metasediments, amphibolites and serpentines predominate. A sample of 'Old Granite' from this assemblage has yielded a radiometric age of 3400 m.y. and some of the supracrustal host-rocks apparently pre-date this granitic material.

Greenstone belts occupy several clearly defined east-north-east tracts within which the metamorphic grade is low and the primary structures are well-preserved. The supracrustal rocks of these belts, constituting the *Swaziland System*, are ideally displayed in the *Barberton Mountain Land* on the borders of Swaziland and the Transvaal where the rugged terrain provides a three-dimensional view of their structure (Fig. 6.3). The brothers M. J. and R. P. Viljoen, and C. R. Anhaeusser have provided accounts which now serve as a basis for comparison with similar formations elsewhere.

The Swaziland system of the Barberton area has a minimum age of 3000 m.y. attested by isochrons for pegmatites emplaced in it and for shales from its middle (Fig Tree) division. It may be significantly older, since lead model ages for galena in veins give dates of nearly 3500 m.y., and provisional ages of 3360 m.y. have been obtained for acid rocks near the top of the lower division.

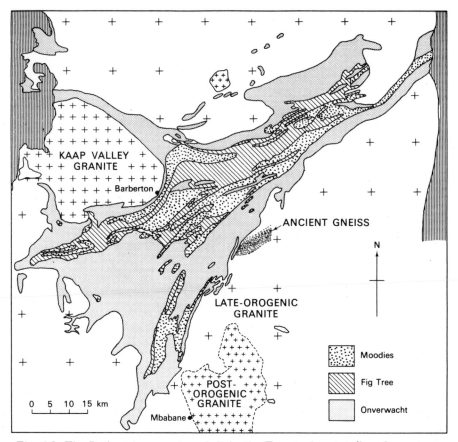

Fig. 6.3. The Barberton greenstone belt in the Transvaal craton (based on maps of Anhaeusser, Viljoen and Viljoen)

It reaches aggregate thicknesses of over 15 km, of which much more than half is volcanic (Table 6.2). Volcanic rocks predominate in the thick Onverwacht Series which forms the basal member of the succession and which includes pillow-lavas, serpentines and cherts in an assemblage of ophiolitic type. The basic lavas are usually pillowy (Plate III) and are associated with important ultramafic components (some also pillowy) towards the base, and with intermediate and acid lavas or tuffs, especially towards the top. The ultramafic and associated basic rocks of the lower portion of the succession have certain chemical peculiarities which have led Anhaeusser and the Viljoens to group them under the name *komatiites* – they are unusually rich in magnesia and poor in alumina and potash. The basic rocks at higher levels more closely resemble ordinary basalts.

Throughout the volcanic sequence a remarkable form of cyclic vulcanicity is apparent in the repetition of a sequence leading from more basic types to thin developments of more acid differentiates and often terminating with a thin chert. The chert horizons, only a few metres thick but quite persistent, are

Table 6.2. A GREENSTONE BELT SUCCESSION:
the Swaziland System of the Barberton Mountain Land
(based on publications of Anhaeusser, Viljoen and Viljoen)

Moodies Series	polymict conglomerates, feldspathic sandstones, siltstones, shales (>3500 m)
(local unconformity)	————————————————————————————————→
Fig Tree Series	3: greywackes and greywacke-shales (banded cherts, iron-formations, tuffs) 2: greywackes, greywacke-shales, banded cherts. 1: mainly chemical sediments (banded cherts, talc-carbonate rock, quartz-sericite rock) (total thickness of Fig Tree Series >3000 m)
Onverwacht Series	3: Upper Onverwacht (Houggenoeg stage): cyclic repetitions of pillow basalts or andesites, acid lavas and cherts, followed by acid lavas, pyroclastics and volcanogenic sediments (5000 m). 2: Middle Onverwacht (Komati River stage): altered pillow basalts and ultrabasic lavas or shallow intrusives, with feldspar-porphyry intrusives (3−4000 m). 1: Lower Onverwacht (Theespruit stage): basic pillow lavas and ultrabasic lenses, containing thin intercalations of black chert, other siliceous sediments and acid tuffs.

banded, often carbonaceous and sometimes ferruginous; they appear to mark periods of quiescence prior to the onset of a new cycle of vulcanicity. The cyclic repetition of igneous types suggests the repeated introduction and differentiation of new magma. It is superimposed on chemical variations of another order of magnitude, expressed by the transition from predominantly ultramafic-basic groups in the lower divisions to predominantly intermediate to acid groups at the top of the Onverwacht Series. Analogies may be seen with the chemical variations in Canadian greenstone belts (p. 68) and with the successions of certain volcanic island arcs.

The sedimentary divisions which overlie the Onverwacht Series are largely detrital and immature in character, though cherts resembling those of the volcanic cycles recur at many levels, especially near the base. The Fig Tree Series consists mainly of fine-grained turbidites derived from both volcanic and granitic sourcelands. The Moodies Series, generally coarser-grained and cross-bedded, carries material derived from the underlying volcanics as well as abundant fragments of granites and alkali-feldspars presumed to be derived from the basement. Both divisions suggest rapid accumulation in unstable troughs close to mountain terrains. A remarkable feature which will be referred to in Chapter 10 is the occurrence, in certain of the cherts, of organised bodies regarded as microfossils and of complex hydrocarbons.

The supracrustal rocks form a broadly synclinal structure projecting like the keel of a boat into the enveloping sea of granites and gneisses (Fig. 6.3). The outer contact is generally made by rocks of the Onverwacht Series facing inward and, consequently, the granitic rocks lie structurally and stratigraphically in the position where a basement would be expected. It is, however, very clear that the

Plate III. Pillow-lavas of the Onverwacht Group, illustrating the low degree of deformation and the preservation of primary structures. Barberton Mountainland, Transvaal

neighbouring granites are intrusive into the base of the Swaziland Series. A sharp increase in grade of metamorphism is seen near the contacts; the adjacent granites tend to be homogeneous and may vein the lavas or carry swarms of fragments derived from them. In detail, the contacts are moulded on the walls of a number of granitic domes which bulge up into the supracrustal series and are separated by screens of highly altered volcanic rocks (Fig. 6.3). In the interior of the greenstone belt, the higher members of the succession are crowded in tight, complex synclines, often separated from one another by dislocations. The metamorphic grade is very low and primary structures may be preserved without apparent distortion. The rapid inward fall in metamorphic grade suggests the influence of a rather steep geothermal gradient, a possibility which is reinforced by the occurrence of andalusite in some pelitic metasediments. Gold-bearing sulphide deposits and quartz-veins — the first gold ores to be worked in southern Africa — are located along fracture-systems formed in the course of regional deformation; many authorities regard the gold as a product of basic vulcanicity, mobilised and concentrated by the action of the granites.

From the relationships here outlined it may be inferred that the broad granitic terrains which surround the greenstone belt include the regenerated derivatives of a pre-Swaziland basement; such a basement has been recognised by Hunter to the south of Barberton (1970). The various granite domes and sheets which cluster around the belt clearly reached a highly mobile state in which they were capable of intruding the supracrustals; some of these bodies have given radiometric ages of about 3000 m.y. The evidence for the existence of a basement of continental type (i.e. of granitic composition) between and at least locally beneath the volcanic pile, makes us reluctant to press too far the comparison with island arc environments of more recent geological times, since island arc volcanics generally rest on oceanic crust. We therefore prefer to envisage a situation such as that outlined by Anhaeusser and others (1969) in which greenstone belt vulcanicity was initiated along flaws in a thin unstable granitic crust, which was subsequently thickened by *underplating*, that is, by the addition of newly differentiated granitic material from below (Fig. 6.4).

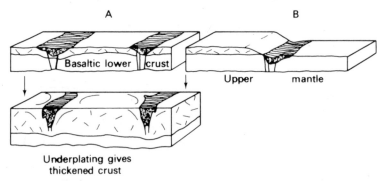

Fig. 6.4. Possible environments of greenstone belt formation: (A) greenstone belts developed along lineaments in thin granitic crust; (B) greenstone-belts located at a primitive continental margin (based on Anhaeusser *et al.*, 1969)

V The Earliest Cratonic Cover of the Transvaal Province

1 The Succession

Although some of the younger granites may have been emplaced as late as 2600 m.y., it seems that the process of stabilisation set in remarkably early in the Transvaal province. Most of the granites were emplaced by 3000 m.y. and by about 2800 m.y. the province had been intruded by basic dykes and planed off by erosion to receive the first division of an extensive cratonic cover-succession. This succession is outlined in Table 6.3, on which the relevant geochronological evidence is given, and its distribution is shown in Fig. 6.5. It would be difficult to exaggerate the geological and economic interest of this long succession. Each of the main divisions, the Witwatersrand and the Transvaal, reaches a maximum thickness of more than 8 km; the rocks of which they are composed have suffered so little modification that it is possible to investigate variations of thickness and facies and to arrive at some understanding of the palaeogeography. The sedimentary assemblages are made up of shallow-water or non-marine sediments, usually of well-sorted detrital or chemical types. The lavas and igneous intrusives are of kinds typical of stable crustal environments. The cratonic cover as a whole shows low dips, and severe structural disturbances are localised, the angular discordancies which separate successive members being slight. The Witwatersrand System is the world's most productive source of gold. The Transvaal formation contains important iron ores and the Bushveld igneous complex carries deposits of platinum, chromium, titanium—iron and tin. A substantial part of the mineral wealth of South Africa is concentrated in this very ancient cratonic sequence.

A minimum age-limit for the base of the succession is provided by a K–Ar date of 2700 m.y. obtained from a rhyolite of the *Dominion Reef Series* which locally underlies the Witwatersrand System. Lead isotopic data suggest that a more realistic figure is 2800 m.y. This impersistent group, consisting principally of acid volcanics, is considered to rest unconformably on the crystalline complex and therefore lies at the base of the cratonic cover. At a much higher level in the succession, the Bushveld igneous complex, which intrudes the Transvaal System, has given a date of 2050 m.y. The main part of the cratonic cover was therefore deposited between 2800 and 2000 m.y.

Table 6.3. THE ARCHAEAN CRATONIC SUCCESSION IN THE TRANSVAAL PROVINCE

million years		
<1790	Waterberg and Loskop systems	principally red or brown sandstones conglomerates, shales: range unknown, some possibly late Precambrian
		←———————————— (*unconformity*) ————————————→
2050	(Intrusion of *Bushveld igneous complex c.* 2050 m.y.)	
	Transvaal system	quartzites, dolomites, banded ironstones, shales, andesites, felsites
		←———————————— (*local unconformity*) ————————————→
2300	Ventersdorp system	basic and intermediate volcanics, conglomerates, sandstones
		←———————————— (*local unconformity*) ————————————→
	Witwatersrand system	quartzites, shales, conglomerates: locally underlain by Dominion Reef Series
2800		rhyolites, *c.* 2800 m.y.
		←———————————— (*major unconformity*) ————————————→
3000–3400	Crystalline basement	Swaziland System and associated granites and migmatites.

2 *The Witwatersrand System*

The best-known outcrops of the Witwatersrand are those of the Rand itself on which the goldfields of the Johannesburg region are situated. From this region, the formation extends beneath a younger cover over a wide area to the south-west and south-east. It is of similar facies throughout and consists

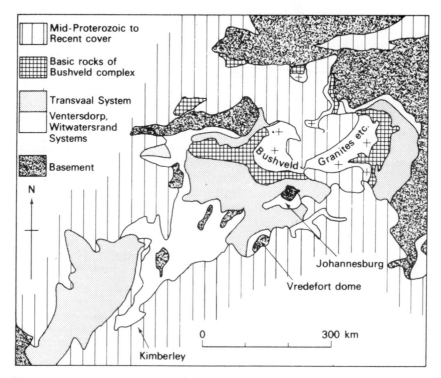

Fig. 6.5. Simple map of the Transvaal region showing basement, Transvaal and Witwatersrand systems and Bushveld complex

essentially of alternations of quartzites and shales or slates with thin quartz-pebble conglomerates. These conglomerates or *bankets* carry gold and uranium and provide the economic ores.

The system is thickest on the Rand where it reaches a total of about 8 km. Several sedimentological features suggest that it was derived from source-regions to the north-west. Most of its sub-divisions thin irregularly towards the south-west or south-east. The pebble-size in the conglomerates decreases south-eastward, the orientation of current bedding in certain quartzite horizons suggests south-eastward flow of palaeocurrents and 'pay-streaks' (streaks abnormally rich in gold which often occupy small erosion-channels) fan out towards the south-east. In the concealed goldfields of the Orange Free State, there are indications of an approach to the south-western margin of a basin of deposition and it is supposed that the original extent of the formation was not much greater than its present extent. Several dome-like inliers of basement within, and on the borders of, the Witwatersrand basin appear to mark the sites of positive areas which supplied detritus during the period of deposition or were subsequently elevated to bring about marginal tilting of the formation.

The Witwatersrand system falls into two sub-divisions: in the *Lower Witwatersrand* quartzites alternate with shales and pebble-bands are uncommon;

the *Upper Witwatersrand* is predominantly psammitic and contains many thin conglomerates. These gold-bearing bankets or *reefs* though seldom more than a few metres in thickness are surprisingly persistent and have received individual names such as Leader Reef, Main Reef, Bird Reef and so on. Their pebbles, up to about 5 cm in diameter, are predominantly of vein quartz, with less common quartzite, chert, jasper and acid volcanics. Up to about 12 per cent of the matrix consists of pyrite, much of it clearly of replacement origin. The gold, carrying some 10 per cent of silver, is largely associated with this pyrite.

Origin of the Witwatersrand deposits. The Witwatersrand sediments are predominantly detrital deposits laid down in the shallow water of a flood-plain or shallow sea. The sediments were derived mainly from the north-west, and it is reasonable to suppose that the source-rocks were those of the crystalline basement which, as we have seen, is itself gold-bearing. The simplest explanation of the origin of the gold-deposits – consistent with their distribution and their association with the coarsest sediments – is that they are essentially ancient *placers*. Some geologists, notably Harwood and Graton, have favoured a hydrothermal origin for the gold, regarding it as being introduced into conglomeratic host-rocks along with the associated pyrite. Davidson (1965) attributed the formation of the reefs to the action of groundwater concentrating disseminated gold in the most permeable horizons. While the textural evidence suggests that there was a good deal of redistribution after deposition, we prefer to regard the gold as syngenetic and the reefs as modified placers.

3 The Ventersdorp system

The Ventersdorp system follows conformably on the Witwatersrand in the centre of the basin, but elsewhere overlaps onto lower groups. It is mainly a volcanic formation which is seen intermittently over an area of 250 000 km^2 and reaches thicknesses of over 1.5 km. The Ventersdorp consists of altered basaltic or basaltic–andesitic lavas (often amygdaloidal) associated with pyroclastic material, acid lavas, conglomerates, feldspathic sandstones and occasional cherts and shales. The assemblage bears some resemblance to that of the plateau-basalts of later times.

4 The Transvaal system

The thick Transvaal system which follows the Ventersdorp is even more extensive, cropping out at intervals over the Transvaal and Griqualand West in a belt which measures over 1000 km from ENE to WSW (Fig. 6.5). It attains its maximum thickness of about 8 km in the Bushveld region. The succession rests on a plane of marine erosion and is in large part marine, possibly passing up into estuarine or lacustrine deposits. A basal division of dark quartzites and pebble-conglomerates, sometimes auriferous (the *Black Reef Series*), is followed first by a thick *Dolomite Series* and finally by the *Pretoria Series* of shales, sandstones and cherts with intercalated andesitic and felsitic lavas. The Dolomite

Series is locally oolitic and exhibits stromatolite structures. It carries chert layers and in many places passes up into cherty quartzites and a banded iron formation which reaches 300 m in the Cape Province and is the principal source of hematitic iron ores in the system. The Pretoria Series includes a bouldery horizon interpreted as a tillite. A thick mass of *Rooiberg felsites* apparently forms the highest division of the Transvaal system.

5 Basic intrusions in the cratonic succession

To the north of Pretoria, the lower parts of the Transvaal system contain many basic sills and are overlain by the huge basic sheets, approaching 10 km in thickness, which constitute the *Bushveld igneous complex* (Fig. 6.6). Outcrops

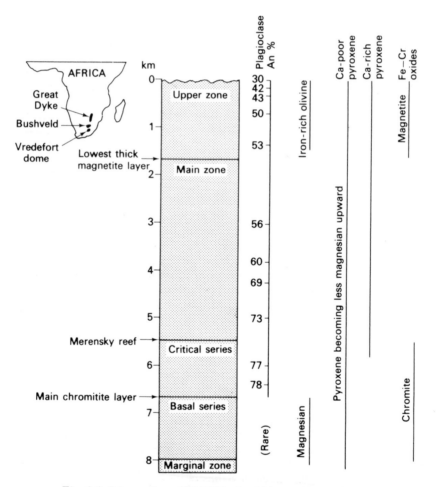

Fig. 6.6. Schematic section of Bushveld igneous complex

of the complex form a broken ellipse which has for long been regarded as a single huge stratified lopolith floored by the Pretoria Series and roofed by Rooiberg felsites; alternatively, it may consist of two or more sheets emplaced separately at about the same structural level, or of a funnel-shaped intrusion.

The outstanding features of the Bushveld complex are, first its size and second its remarkable igneous layering due mainly to crystallization-differentiation *in situ*. Igneous cumulates are represented by peridotites, pyroxenites with chromite layers, banded hypersthene-rich gabbros and anorthosites. Remarkable concentrations of platinum occur in one thin horizon (the Merensky reef) and/or magnetite—ilmenite in others. Above the stratified basic and ultrabasic assemblage lies an almost equal thickness of 'Red Granite' (an assemblage of granite, granophyre, granodiorite etc.) overlain by ignimbrites and acid lavas forming the 'Rooiberg felsites'. These voluminous extrusive and near-surface acid rocks are tentatively regarded as derivatives of crustal material melted by the ascending basic magma.

The Bushveld complex, large though it is, appears to be only one of a string of basic complexes emplaced along a northerly lineament in the period 2500—2000 m.y. The Great Dyke of Rhodesia (p. 113) follows this lineament for 500 km and has been interpreted as an intrusion emplaced in a tensional rift formed by sub-crustal convection beneath the Archaean craton of Rhodesia. Two buried bodies on the southward projection of the lineament are indicated by positive bouguer anomalies. One underlies the *Vredefort dome* of the Transvaal — an extraordinary structure defined at the surface by an upthrust plug of granitic basement ringed by a deformed and hornfelsed collar of Witwatersrand and Transvaal sediments — and the other lies some 100—200 km further south.

A somewhat later igneous event, of interest in the light of the later history of southern Africa, was the emplacement of the diamond-bearing kimberlite pipe which is worked in the Premier Mine north of Pretoria. Although most African kimberlites are of Mesozoic age, that of the Premier Mine has yielded a galena age of over 1700 m.y. and is securely established as more than 1115 m.y. by the dating of an intrusive dolerite. This evidence indicates that kimberlite emplacement, so characteristic of stabilised cratons, had begun in southern Africa in mid-Proterozoic times.

VI The Rhodesian Province

The basement of the Archaean province of Rhodesia is very similar to that of the Transvaal from which it is separated by the Limpopo belt. It is made up of the same two assemblages — an association of granodioritic rocks, gneisses and migmatites forming domes and broad irregular outcrops, and a supracrustal greenstone assemblage forming narrower tracts known locally as 'schist belts' or 'gold belts' (Fig. 6.7). The schist belts are broadly synformal in structure; the underlying granitic complexes include granitic material which clearly intrudes the supracrustal rocks but which may represent the product of reactivation of basement rocks.

The schist belts exhibit a rather consistent succession in which two divisions incorporating abundant basic volcanic material are followed by an upper division consisting largely of badly-sorted clastic sediments (Table 6.4). *The Sebakwian Group* is almost invariably located along the outer margins of the synclinal schist belts. Its metamorphic grade is generally higher than that of the overlying divisions, its rocks are frequently migmatitic and its base is swamped by invasive granitic material. The dating at nearly 3400 m.y. of veins which cut migmatitic host-rocks suggests a very early date for accumulation.

The *Bulawayan Group* contains the major part of the volcanic assemblages of the schist belts and reaches thicknesses of up to 10 km. The Bulawayan volcanics, the principal host-rocks of the Rhodesian gold mineralisation, are strikingly like those of the Onverwacht Group of Barberton. Sedimentary intercalations in the Que Que area include feldspathic sandstones and polymict conglomerates carrying granite pebbles which are thought to be derived from the neighbouring Que Que granite. K–Ar dates for these pebbles are in the region of 3400 m.y. A limestone horizon north of Bulawayo carries well-preserved stromatolites, the oldest known record of algal-mat formation. *The Shamvaian Group*, the youngest division of the schist belts, is predominantly sedimentary and has some of the lithological features of molasse. Although evidently laid down after the emplacement of many granites, the group is deeply infolded in the schist belts and is cut by younger granites. It therefore cannot be regarded as a post-orogenic formation of true molasse type.

Towards the western side of the Rhodesian massif, the characteristic pattern of synformal schist belts pinched between adjacent domes of granitic material is clearly revealed (Fig. 6.7). Towards the east, the schist belts are reduced to a few screens and tonalites, granodiorites and granites form a vast territory within

Table 6.4. THE ARCHAEAN SUCCESSION OF THE RHODESIAN SCHIST BELTS

Shamvaian Group	Predominantly sedimentary, feldspathic psammites, polymict conglomerates, greywackes, shales, rare limestones, banded iron formations and volcanics: folded Shamvaian cut by pegmatites dated 2650 m.y.
(unconformity) ————————————————————————→	
Bulawayan Group	Predominantly volcanic, basaltic pillow-lavas, andesites, pyroclastics, rhyolites, greywackes, shales, banded iron formations, limestones, intrusive ultramafic and mafic sheets; probably older than granites and pegmatites dated about 3000–2900 m.y.
(unconformity) ————————————————————————→	
Sebakwian Group	Psammites, ultramafic lavas and intrusives, basic volcanics preserved partly as remnants in migmatites: cut by veins dated at 3300 m.y.

(no base preserved: pre-Sebakwian supracrustal groups incorporated in gneisses and granites underlying the schist belts).

which some individual domes are indicated by a concentric arrangement of the foliation. MacGregor used the expression 'gregarious batholiths' to describe the clustered domes whose average diameter is 60–70 km. In a far-seeing synthesis (1951), he outlined an interpretation of their relationships in terms of the reactivation of granitic basement material. As we have noted, the schist belts are broadly synformal — some form 'triangular synclines' in the angle between three

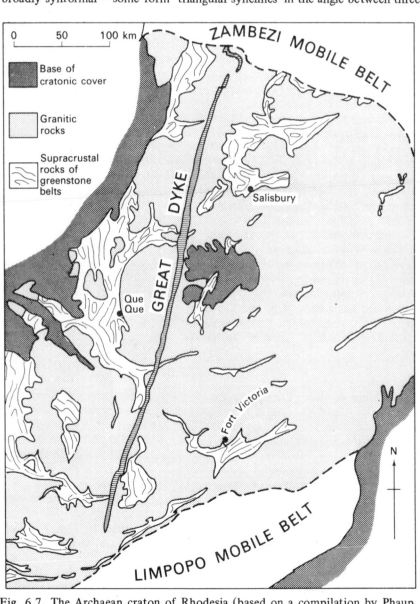

Fig. 6.7. The Archaean craton of Rhodesia (based on a compilation by Phaup, 1971)

domes – and in general terms their successions rest on, and young away from, the granitic complexes. The low density of granitic materials may be expected to favour the development of domes under temperature conditions in which these materials become mobile. Talbot (1968) attributes dome formation to thermal convection, at temperatures well below those required for melting, in a granodioritic crustal layer some 30–40 km in thickness.

These satisfyingly simple concepts require some qualification. The granites of the craton are known, both from isotopic dating and from their relationships with the successive groups of the schist belts, to have been emplaced over a long time-period, ranging at least from 3300–2600 m.y. Gold mineralisation, mostly in the form of quartz-sulphide veining, appears to have been associated with more than one episode of granite-formation and lithium pegmatites, occasionally associated with tin, accompany late (2600 m.y.) intrusions. Periods of folding and metamorphism intervened between the deposition of the three supracrustal groups, and some of the complex structures formed in the course of these episodes appear to have had no connection with the processes of dome formation. In the Selukwe region, for example, Stowe (1968) has recognised nappe-structures in which inverted supracrustal successions were subsequently distorted by updoming. Finally, some of the granites of the massif are clearly intrusive and are associated with thermal aureoles in the schist belts.

Taking all these features into account, one must conclude that the simple looking dome-pattern of the Rhodesian massif is the end-product of a long sequence of crustal movements. The occurrence of granitic debris in clastic rocks from each of the three stratigraphical divisions suggests that the granitic crustal material was repeatedly elevated and subject to erosion. To us, it seems possible that a contrast between the dome-areas and the intervening 'sinks' was established at an early stage and thereafter exerted an influence on both the pattern of deposition and the pattern of deformation – the buoyancy of the granitic areas being expressed in various ways according to the crustal conditions.

The Limpopo belt. The east-north-east Limpopo belt which separates the Archaean massifs of Rhodesia and the Transvaal has a characteristic metamorphic and tectonic style. The rocks incorporated in it appear to include representatives of the schist belt granitic assemblages together with anorthosites and shallow-water sediments, all showing the overprint of metamorphism of amphibolite or granulite facies. Large-scale interference-patterns recording the effects of repeated plastic deformation are traversed by 'straight belts' characterised by flaser gneisses and platy rocks. These features suggest that the Limpopo belt remained a zone of high temperatures and stresses after stabilisation of the adjacent massifs. Isotopic determinations by several methods have given ages of about 2000 m.y., which provide a minimum date for the end of the 'Limpopo cycle'. Since satellitic intrusions of the Great Dyke enter the belt without apparent distortion or metamorphism, it must be presumed that effective orogenic activity came to an end in some regions before intrusion of these bodies at 2500 m.y.

The Great Dyke

The Great Dyke of Rhodesia, which may be regarded as an early manifestation of the cratonic basic magmatic activity already mentioned (p. 108), follows a north-north-east lineament for nearly 500 km (Fig. 6.7). It is a straight-edged body averaging nearly 6 km in width and cutting indifferently across the Archaean structures. It is not a 'dyke' in the sense of an intrusion extending vertically downward to great depths, but may represent the product of intrusion into a graben. The ultrabasic and basic rocks of which it is made form a string of layered complexes, each of which may have been fed from a separate point along the lineament. The igneous stratification, dipping inward from the walls, defines a sequence of ultrabasic units followed upward by gabbros. Serpentinised dunites and pyroxenites with seams of chromite form the greater part of each complex and are thought to represent the accumulated products of repeated intrusion; each new cycle begins with a chromite seam followed by olivine-rich and then by pyroxene-rich rocks. Layered gabbros are preserved as boat-shaped outcrops in the highest parts of each complex.

VII The Tanzania Province

The Archaean province of central Tanzania and southern Kenya forms a unit very similar to the Rhodesian province, in which the principal assemblages are granitic complexes on the one hand and supracrustal belts on the other. Extensions of the province north-westward into Uganda, Congo (Leopoldville) and the Central African Republic show the overprint of Proterozoic phases of mobility, but will be referred to briefly because they provide evidence of early structures not preserved in the main massif. Details of the Archaean components are summarised in Table 6.5.

Basement gneisses unconformably underlying supracrustal rocks of green-stone belt type are recognisable in the north-western salient where charnockites and a variety of gneiss complexes are unconformably followed by the Kibalian and Buganda-Toro groups. With the exception of a lead model age of 3480 m.y. for galena, the few age determinations from this little-known terrain reflect post-Archaean phases of activity. Nevertheless, it is reasonable to think that a granitic crust dating from about 3400 m.y. is preserved here.

The rocks of the supracrustal belts of the Tanzanian massif define an irregular east-south-east grain. In the northern part of the massif they include a lower division of volcanics with associated cherts, banded ironstones and black pelites (the Nyanzian and the Buganda-Toro and Kibalian, regarded by Shackleton as equivalent groups) and an upper division of immature clastic sediments carrying both volcanic and granitic debris (the Kavirondian). In these northern regions the supracrustal belts are clearly defined and only slightly metamorphosed; they are folded on steeply dipping axial planes and are invaded by granites of at least two generations.

In the southern part of the massif, supracrustal rocks similar in lithology to those of the Nyanzian occur mainly as enclaves of amphibolite facies enclosed in

Table 6.5. THE TANZANIA PROVINCE

Archaean units of the massif

Post-Kavirondian granites *c.* 2600 m.y.	
	Kavirondian System Poorly sorted clastic sediments, polymict conglomerates, feldspathic sandstones, shales

←———————————— *(unconformity)* ————————————→

Granites and pegmatites *c.* 2900 m.y.	*Nyanzian System* Basic volcanics, with acid volcanics, black pelites, cherts, banded iron formations possibly equivalent to the *Dodoman* (Tanzania) and the *Buganda-Toro* and *Kibalian* (Uganda))

←———————————— *(?major unconformity)* ————————————→

? 3480 m.y.	*Basement gneisses* Grey gneisses of Congo (Leopoldville), and charnockitic and granitic assemblages of Uganda (West Nile Series, Watian, Aruan etc.)

granites of migmatitic aspect. These remnants appear to represent the resistant portions of a migmatised series. They have been assigned to a 'Dodoman' system thought to be older than the Nyanzian, but are probably the equivalents of the Nyanzian digested during the reactivation of basement granites. The characteristic gold-quartz mineralisation of the massif is localised in the more northerly, higher-level regions around Lake Victoria, where veins occur in shear-zones in the supracrustals close to their contacts with the granites.

VIII Early and Middle Proterozoic Belts of Eastern and Southern Africa

Several tectonic provinces attained their structural form during periods of mobility which followed the stabilisation of the Archaean massifs. Setting aside some little-known units, we may concentrate on two early Proterozoic belts (the *Ubendian-Rusizian belt* of western Tanzania and adjacent parts of the Congo and the *Kibali-Toro belt* of Uganda) as well as one long mid-Proterozoic belt: the *Karagwe-Ankole belt* of East Africa and its equivalents in South Africa. The late Proterozoic network is left for Part II.

1 The Ubendian-Rusizian belt (1800 m.y.)

The Ubendian belt cuts obliquely across the tectonic grain of the Tanzania Archaean province and follows a south-easterly course for about 1000 km (Fig. 6.8). It is not more than 200 km in width and appears to have been developed between two stable forelands – the Tanzania massif on the north-east and the Kasai massif, now largely obscured, on the south-west. The southern parts of the Ubendian belt are made largely of gneisses of amphibolite or granulite facies, among which charnockites have been recorded. Relics of high-grade pelites, quartzites and marbles are abundant. On the western side of Lake Tanganyika, a partially migmatised sequence of schistose or phyllitic pelites, quartzites and conglomerates constitutes the *Rusizian*.

The Ubendian belt is characterised by a strong north-westerly grain, parallel to the margins of the belt, which is followed by zones of shearing and by narrow zones of migmatites. The late-tectonic Kate Granite of Southern Tanganyika, following the same trend, is some 200 km in length and only about 15 km in width. A curious structural feature of the Ubendian belt is the occurrence, in several regions, of steep zones of mylonitisation or cataclasis, often parallel to the tectonic trend but sometimes forming conjugate sets.

2 The Toro belt

In Uganda and Northern Congo a rather narrow zone of west-north-west trend contains the *Toro, Buganda* and *Kibalian* groups which consist mainly of pelitic and quartzitic metasediments with cherts and volcanics. Minerals from pegmatites and granites associated with those groups give ages within a fairly restricted range of 2000–1700 m.y. This consistency, with a certain continuity of structural and metamorphic patterns, has led to the recognition of a *Toro orogenic belt* broadly contemporaneous with the Ubendian belt described above. There is, however, some reason to suppose that the supracrustal groups incorporated in the belt are Archaean (see Table 6.5) and the status of the belt is doubtful. Neither the Toro nor the Ubendian belt is associated with a distinctive suite of mineral deposits.

3 The Karagwe-Ankole belt and its equivalents

One of the most distinctive orogenic belts of Africa is the Karagwe-Ankole belt which extends south-westward for at least 1500 km from Uganda to Zambia. The name *Karagwe-Ankole* applied to this belt in the literature of former British territories is replaced in that of former Belgian territories by the name *Kibaran belt*. In South-West and South Africa, the *Orange River belt* which loops around the southern side of the Transvaal massif, forms a tectonic province perhaps originally continuous with the Karagwe-Ankole belt. Isotopic dates for late-tectonic granites and pegmatites in both provinces fall in the range 1050–900 m.y., giving a firm date for the terminal stages of the geological cycle. The belt as a whole is clearly younger than the end of the Ubendian cycle (c. 1800 m.y.) and the dating of early granites at about 1300 m.y. suggests that it had a life-span of several hundred million years.

In central Africa, the Karagwe-Ankole belt obliquely crosses the Ubendian belt, portions of which are seen across the full width of the younger structure in a swell from which most of the cover has been removed (Fig. 6.8). Where it is

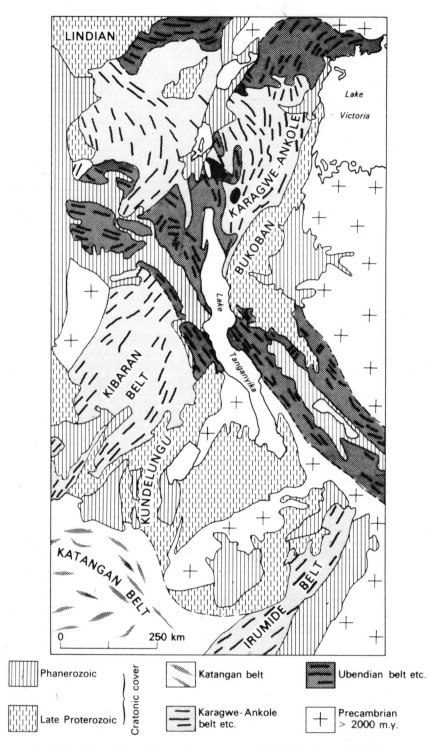

LINDIAN

Lake
Victoria

KARAGWE-ANKOLE

BUKOBAN

Lake
Tanganyika

KIBARAN
BELT

KUNDELUNGU

KATANGAN BELT

IRUMIDE BELT

0   250 km

	Phanerozoic	Cratonic cover		Katangan belt		Ubendian belt etc.
	Late Proterozoic			Karagwe-Ankole belt etc.		Precambrian > 2000 m.y.

Fig. 6.8. The relationships of successive mobile belts in central Africa

clear of the Ubendian province, the belt is flanked by Archaean massifs to east and west and similar massifs border the Orange River belt in the south. The apparent continuity of the Ubendian belt through the Karagwe-Ankole mobile zone suggests that the younger mobile belt was developed within a cratonic continental mass rather than at a former continental margin.

Within the Karagwe-Ankole province of central Africa, the characters of the cover-series, the tectonic and metamorphic style, and the mineralisation remain fairly consistent. The cover-succession is known in Katanga as the *Kibaran*, in Burundi as the *Burundian* (originally 'Urundian') and in Uganda as the *Karagwe-Ankolean*. It consists predominantly of psammitic and pelitic metasediments with minor calcareous horizons and conglomerates and local groups of volcanics. Unconformities divide the succession in some regions and there is persistent evidence of shallow-water deposition. Near the northern extremity of the belt, south of Lake George, the proportion of coarse sediment increases northward and bodies of conglomerate and quartzite have been interpreted as delta-fans laid down by rivers flowing off the foreland. In thickness (6–11 km), the succession is typical of mobile belts, but it is somewhat unusual in the scarcity of turbidites and in the fact that volcanic material plays a very subordinate role.

In the main Karagwe-Ankole belt, recumbent folds or other structures indicative of strong crustal shortening are scarcely represented; major thrusts are reported in Katanga but are of little importance elsewhere; in south-west Uganda, for example, folding dies out gradually towards the eastern foreland and there is no thrust-front. In Katanga, the fold-structures have steep axial planes roughly parallel to the trend of the fold-belt, but in Burundi and Uganda two sets with axial planes almost at right angles produce complex interference-patterns. The complexity of the fold-system and of related minor structures increases with depth, in harmony with an increase of metamorphic grade (see below). Repeated deformation is indicated by refolding and by the superposition of successive sets of minor structures.

In the regions of complex folding, sharp culminations at the meeting-points of anticlinal axes produce steep-sided domes in which granites and gneisses may be revealed. These domes often form low-lying areas, known in Uganda as 'arenas', encircled by hills of Karagwe-Ankolean quartzites. The granitic bodies which occupy many of the arenas are demonstrably intrusive and late-tectonic, and have yielded isotopic ages of about 1000 m.y. Nevertheless, they lie consistently below the cover-succession and thus, as King has put it, 'the expected position of the unconformable base of the sediments is typically represented by an intrusive junction with younger granite' (1959, p. 740). This anomaly recalls the relationships of mantled gneiss domes and it seems very probable that the granites and gneisses of the arenas represent the partially mobilised basement.

The metamorphic zones of the Kagarwe-Ankole belt in central Africa show a remarkable constancy of arrangement – the grade increases with depth and the isograds are roughly parallel to the stratigraphical boundaries. Most of the cover-succession shows a low grade of metamorphism and the highest groups are virtually unmetamorphosed. Towards the base of the succession, pelitic rocks may change from slates or phyllites to mica-schists in which kyanite, staurolite, andalusite or cordierite appear. This downward passage to schists or even to migmatitic gneisses takes place at a higher stratigraphical level in the central parts of the belt than it does near the margins and on a broad scale, therefore, the isograds rise towards the interior of the belt. On a smaller scale, they reflect the fold-pattern and are concentrically arranged to define aureoles about the granites and gneisses of the arenas. The schist zone is seldom more than about 2 km in thickness.

The distinctive *mineralisation* of the Karagwe-Ankole belt is associated with the numerous late-tectonic granites. The granites themselves carry small amounts of tin, while tin, tungsten, niobium, tantalum and lithium occur in pegmatites and quartz veins emplaced in the contact-rocks. The consistent style of mineralisation along the length of the belt provides a striking illustration of the idiosyncratic nature of orogenic cycles. There is, as we have seen, a good deal of evidence that some, at least, of the granites represent remobilised basement. The basement of the Karagwe-Ankole belt includes both Archaean and early Proterozoic complexes none of which are characterised by tin mineralisation. In this instance, it would seem that the elements of the ore-deposits must have been introduced by some mechanism related to the development of the Karagwe-Ankole mobile belt.

Post-tectonic sediments of molasse type associated with the Karagwe-Ankole belt occur in two main basins flanking the fold-belt, as well as in smaller intramontane basins. These sediments include boulder-conglomerates, arkoses and feldspathic sandstones sometimes associated with amygdaloidal basalts reaching thicknesses of up to 2500 metres. They are followed, in many places with slight unconformity, by 'platform' sediments which are equivalent to the main cover-succession in the late Proterozoic Katangan mobile belt. The molasse-deposits, laid down rapidly after the ending of orogeny in the Karagwe-Ankole belt, form the lower parts of the *Bushimay, Malagarasy* and *Bukoban* groups (Table 6.6).

The *Orange River belt* of southern Africa contrasts with the Karagwe-Ankole belt proper in several respects. Its supracrustal cover, known as the *Kheis system*, consists largely of basic and acid volcanics with psammites of various types. Although radiometric dates have not so far provided confirmatory evidence, there are indications that the Kheis supracrustals may be far older than the Karagwe-Ankole, perhaps even equivalent to the Swaziland system. The metamorphic grade is consistently high and vast areas consist of migmatitic gneisses, charnockites and other rocks of granulite facies. Although tin is not of much importance, tungsten is mined at several localities.

IX Proterozoic Cratonic Regions in Central Southern Africa

No mention has been made so far of the history of the various cratonic regions which remained stable over part or the whole of the Proterozoic era. Flat-lying and unmetamorphosed formations older than the Karroo have been encountered in many regions, especially near the margins of the Archaean massifs. Some can be traced laterally into successions involved in the folding of orogenic belts of known age and some can be dated isotopically by means of igneous bodies which intrude them, but the evidence is still too scanty to allow of much generalisation. The names, characters and approximate ages of some of the more important groups are summarized in Table 6.6.

Igneous activity in the cratonic areas can be grouped under a number of heads. *Basic dyke swarms* traverse the Archaean provinces, and related sills or lavas are incorporated in some of the older cratonic cover formations. Vail (1970) gives evidence suggesting that dolerites associated with the Waterberg and Umkondo groups (Table 6.6) fall in the range 1950–1750 m.y., while those associated with the Bukoban, fall in the range 1200–950 m.y.; but in the present state of knowledge it would be difficult to ascertain whether these groups are associated with dyke swarms of the dimensions of those which cross the Canadian shield.

A second group of igneous bodies is of *alkaline character*. In southern Africa, plugs or ring-complexes of syenite, nepheline-syenite, trachyte and carbonatite are spread out along a NNW lineament and associated with a NNW swarm of dolerite-syenite dykes in the Transvaal. The *Pilansberg complex*, one of this set, is dated at about 1400 m.y. We have already seen that the earliest kimberlites date from about 1700 m.y. (p. 109): both these very old manifestations of cratonic igneous activity are located in what is perhaps the oldest cratonic massif of the continent.

Finally, activity of an entirely different and rather unusual type is recorded by the *anorthosite complex* of southern Angola, which lies near the south-western margin of the Congo craton. This complex forms a massif 300 km in length emplaced in a terrain of granites and gneisses. Much of it is made up of a

Table 6.6. PROTEROZOIC CRATONIC COVER-FORMATIONS IN CENTRAL AND SOUTHERN AFRICA

Central Africa	(a)	Equivalents of the cover-succession in the Katangan mobile belts (for details see Part II): *Katangan, Kundelungu, West Congolian, Lindian*, parts of *Bushimay, Bukoban, Malagarasian* (<1200 m.y.)
	(b)	Late-orogenic deposits of Karagwe-Ankole mobile belts: lower parts of *Bushimay* (west Katanga, Kasai), *Malagarasian* and *Bukoban* (west Tanzania, Burundi): non-marine conglomerates, arkoses, brown sandstones, shales, amygdaloidal basic lavas (>1000 m.y.)
Southern Africa	(a)	Equivalents of cover-successions in the Katangan mobile belts: *Nama, Malmesbury*
	(b)	Mid-Proterozoic cover on Archaean province of Rhodesia: *Umkondo Group* (eastern area), *Pirawiri* and *Lomagundi Groups* (north-western area): shallow-water shales, sandstones, limestones with basalts and andesites (2000–1750 m.y.)
	(c)	Mid-Proterozoic cover on Archaean province of Transvaal: *Loskop and Waterberg Systems*: mainly non-marine red or brown sandstones, conglomerates and shales (>1700 m.y.).

pale massive anorthosite which passes marginally into mafic or ultramafic types. This association, tentatively dated as at least 1200 m.y. in age, has been interpreted by Kastlin as a product of calcium metasomatism of gneisses, while a younger troctolitic anorthosite carrying titaniferous magnetite appears to be intrusive. Although the indications are that the Angola anorthosite is not older than mid-Proterozoic, it should be noted that this vast body lies in a zone, extending eastward through Africa to other parts of Gondwanaland, which is characterised by anorthosites of greater antiquity.

X The West African Craton

The cratonic nucleus of West Africa extends from the Gulf of Guinea to the Anti-Atlas of Morocco. It is bounded on the east and west by late Proterozoic-early Palaeozoic mobile belts, and on the north by the front of the Mediterranean system of Mesozoic and Tertiary belts (Fig. 6.2). Especially in the Sahara region, the Precambrian is often blanketed by a Phanerozoic cover and this fact, with the very difficult nature of the terrain, has hindered geological investigation. Nevertheless, there are indications that at least two tectonic provinces can be distinguished, the principal components of which are summarised in Table 6.7.

Table 6.7. PRECAMBRIAN UNITS OF THE WEST AFRICAN CRATON

Later Proterozoic cratonic cover-formations

> *Infracambrian tillite group*
> *Voltaian Group* (Ghana) late Precambrian platform sediments.
> *Tarkwaian Group* (Ghana), low-grade detrital metasediments, possibly post-orogenic formation of Eburnian cycle.

The Eburnian province (2200–800 m.y.)

> In *Ghana and Ivory Coast,* the *Birrimian* supracrustal group of low-medium grade pelites, turbidites and basic volcanics, in *Sierra Leone*, the predominantly sedimentary *Marampa* group. All are folded and intruded by a large suite of *Eburnian granites* the oldest of which give dates of 2200 m.y. Some of these granites are associated with gold-quartz reefs.

The oldest provinces (2900–2500 m.y.)

> In *Guinea and Sierra Leone*, small areas of gneisses associated with amphibolites, schists, banded iron formations (*Kambui schists*) of medium to high grade. Pegmatites carry gold, columbite-tantalite and cassiterite.
>
> In *Ivory Coast*, granulites (*Série de Man*).
>
> In *Ghana, Dahomey and Nigeria*, a north–south tract of gneisses (*Dahomeyan*) generally regarded as Archaean but yielding only Proterozoic ages.
>
> In *the Hoggar* (Sahara), *Pharusian* assemblage of low-grade supracrustals.

The *oldest massifs* are made largely of strongly metamorphosed supracrustal assemblages associated with migmatitic gneisses and granites. Remnants of these rocks occur in a number of separate localities and both the dating and the correlation of the components remain uncertain. The broad northerly tract of gneisses known as the *Dahomeyan* in Ghana, Dahomey and Nigeria, for example, has yielded no dates over about 1850 m.y., although the gneisses are customarily regarded as Archaean. In Sierra Leone, the high grade *Kambui schists* are invaded by pegmatites which give $^{207}Pb/^{206}Pb$ ages of 2900 m.y. These pegmatites and associated veins carry not only gold but also columbite-tantalite and cassiterite, a somewhat unusual assemblage for early Archaean rocks.

The Archaean remnants of the West African craton lie within, or at the edge of, a larger province of very late Archaean or very early Proterozoic age. The

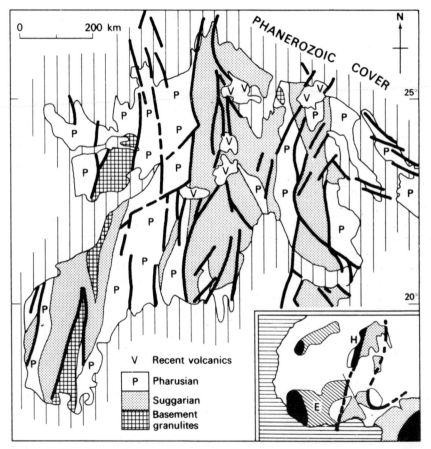

Fig. 6.9. The Precambrian inlier of the Hoggar showing some of the major dislocations (based on Lelubre, 1969). The inset shows the location of the Hoggar (H) and the distribution of Archaean remnants (black tinting), of early Proterozoic provinces (diagonal lines: E = Eburnian) and of late Proterozoic mobile zones (light tinting) in western Africa

principal rocks of this province are the *Birrimian supracrustals* and the large suite of *Eburnian granites* which intrude them. Dated granites fall in the range 2200–1900 m.y., suggesting that the *Eburnian cycle* belonged to the rather localised phases of crustal mobility spanning the Archaean–Proterozoic boundary. We have encountered such phases in the Belomorides of the Baltic shield (p. 23).

The *Birrimian supracrustal sequence* of the type area in Ghana is of eugeosynclinal type, consisting largely of greywackes, greywacke-pelites and basic volcanics. Manganiferous phyllites and gondites (quartz-rhodonite-Mn-garnet rocks see p. 134) are interbedded in the upper groups. Rather similar assemblages in the Upper Volta and Ivory Coast are tentatively correlated with the Birrimian, as are the predominantly metasedimentary Marampa schists of Sierra Leone.

The Birrimian and its equivalents are strongly folded on steep axial planes which have a predominant northerly trend. The grade of metamorphism is variable. In Ghana, pelitic metasediments are represented mainly by phyllites and basic volcanics by greenstones, but further west the common types are schists and amphibolites of higher grade. An enormous bulk of granitic material is associated with the Birrimian formations, probably exceeding them in total area and extending well to the east of the present border of the craton. Much of this material forms concordant granodioritic or monzonitic bodies with broad metamorphic aureoles in which contact-migmatites are sometimes developed. Smaller discordant granites, rich in soda or potash, appear to represent late-tectonic intrusions. Gold in quartz-reefs is associated with late-tectonic post-Birrimian granites in Ghana (hence the old name of the Gold Coast); a reef associated with this mineralisation is dated at about 2200 m.y.

Following the intrusion of most of the granites, late-tectonic sediments were deposited unconformably on the crystalline rocks. The *Tarkwaian* of Ghana is folded on lines roughly parallel to those of the Birrimian. It shows a low grade of metamorphism, is invaded by minor basic and acid intrusions, and may represent a molasse related to the orogenic cycle affecting the Birrimian. It is almost entirely sedimentary and consists of conglomerates, phyllites and quartzites in which detrital gold and diamonds occur.

The deposition of the Tarkwaian in mid-Proterozoic times appears to have marked the final stage before stabilisation in the southern part of the West African craton. Remnants of other Precambrian cratonic cover-formations are preserved locally on the crystalline rocks forming the lowest members of the patchy blanket which includes Infracambrian and Phanerozoic formations.

The Precambrian massif of the Hoggar in the northern part of the craton is separated from the regions already dealt with by a broad expanse of younger sediments. This Saharan massif, isolated in the deserts of Algeria, has been traditionally discussed in terms of two assemblages – the highly metamorphic *Suggarian* in the centre and the east and the low-grade *Pharusian* towards the west. Powerful dislocations and mylonite-zones forming lineaments which can be traced northwards for some hundreds of kilometres separate the outcrops of the divisions.

The relationships of the two divisions are by no means clearly established and contrasts in metamorphic state may well have led to incorrect correlations.

Isotopic age determinations for Suggarian migmatites, schists and granites in the eastern part of the massif have yielded no ages older than about 700 m.y., suggesting that this high-grade terrain belongs to the late Proterozoic system of mobile belts. While most granites in the western part of the massif have given similar dates, one granite, surrounded by low-grade supracrustals assigned to the Pharusian, has given a zircon age of 1900 m.y. and ages of up to 2700 m.y. have been recorded. It seems probable that the western part of the massif belongs to the West African craton and that the spectacular dislocations are connected with movements in the marginal zone of the late Proterozoic belt.

7

The Indian Craton

I Make-up of the Indian Subcontinent

From a geological standpoint, the Indian subcontinent can be separated into two contrasting regions. To the south, bounded by fracture-coastlines, lies the triangular cratonic region of *Peninsular India and Ceylon* which has remained stable since early Palaeozoic times. To the north rises the *Himalayan belt*, a sector of the Alpine-Himalayan mobile belt which has been subject to orogenic disturbance throughout Phanerozoic times. These two regions have had very different histories and it is widely believed that the craton which makes Peninsular India was brought into juxtaposition with the land-mass to the north of the Himalayan belt at a comparatively recent date, as a result of northward drifting after the break-up of Gondwanaland.

II The Indian Craton

Near the Himalayas, the rocks of the Indian craton are submerged beneath debris derived from the erosion of the mountain belt. These Tertiary, Pleistocene and Holocene deposits make the plains of north India and the vast deltas of the Indus, the Brahmaputra and the Ganges. Towards the south, the crystalline foundations are seen beneath an intermittent cover of undisturbed formations ranging in age from late Precambrian to Tertiary.

The traditional classification of the Precambrian rocks of the Peninsula recognised a metamorphosed and much folded 'Archaean' complex separated by a great unconformity – that of the 'Eparchaean Interval' – from overlying unmetamorphosed and undisturbed 'Purana'. With the development of systematic field-studies and the extension of isotopic dating, the old usage of these terms has become meaningless. Instead of one grand orogeny (which Holmes once referred to as a few million years of tectonic hysteria) half-a-dozen cycles have been demonstrated, and undisturbed sediments of appropriate ages have been recognised in association with some of them. For descriptive purposes, we may recognise a number of provinces, or smaller massifs, distinguished

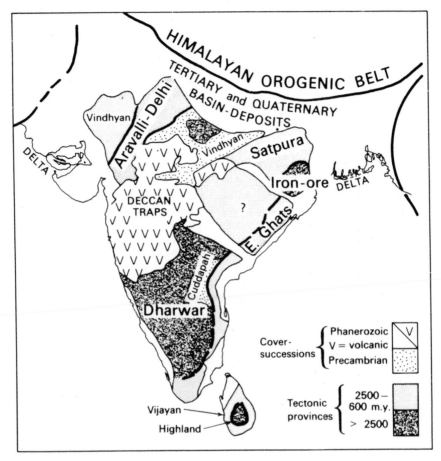

Fig. 7.1. The main geological units of peninsular India

according to the date of stabilisation and arranged in three principal age-groups (Fig. 7.1).

1 The oldest massifs (>2000 m.y.)

Although much of the craton consists of rocks which appear to have originated more than 2000 m.y. ago it is only in relatively small areas that these rocks have been preserved without subsequent modification. The largest of the ancient massifs forms the *Dharwar belt* of south-west India. The gneissose basement in the *Aravalli* region between Bombay and Delhi, and several smaller remnants isolated by outcrops of a cratonic cover are probably of the same general age. A separate massif, bordered by younger fold-belts, makes the *iron ore belt* of the Singbhum area, and smaller, partially reworked massifs exist in south-east India and central Ceylon.

2 The mid-Precambrian belts (2000–1200 m.y.)

The early massifs are truncated by two or three narrow belts in which orogenic mobility appears to have been repeated over a long time-span. The *Eastern Ghats belt* occupying the eastern coastal tract, the *Aravalli–Delhi belt* which runs south-westward from Delhi, and the arcuate *Satpura belt* affected by the *Singbhum cycle* and extending inland from the Calcutta region are the principal units of this group.

3 The late Precambrian belts (600–500 m.y.)

In the south-eastern part of the peninsula and in Ceylon, there are indications of tectonic and thermal events continuing down to about 500 m.y. These episodes, recorded mainly by modifications of much older rocks, constitute the *Indian Ocean* orogeny of Aswathanarayana and will be dealt with in Chapter 3 of Part II.

4 Cratonic supracrustal rocks and intrusives

These rocks, resting on the stabilised crystalline complexes of Peninsular India are, as might be expected, of many different ages. The oldest — parts of the *Cuddapah* in the east and the *Vindhyan* in the north — are now known to date back at least to 1400 m.y.

III The Oldest Massifs

1 The Dharwar belt

The Dharwar belt of southern India is made up of two principal components: an assemblage of metamorphosed supracrustal rocks known as the *Dharwar Series* on the one hand and a variety of gneisses, migmatites and granites on the other. The supracrustal rocks are as a rule arranged in narrow elongated bands sandwiched between broader tracts of gneiss in a manner recalling the association of greenstone belts and granites in other Archaean provinces. The alignment of the supracrustal outcrops defines the north-north-west grain of the province (Fig. 7.3).

In the north, metamorphism in the supracrustal rocks is often only of chlorite grade and original structures such as lava-pillows may be easily recognised. Towards the south-east the metamorphic grade rises and such minerals as cordierite, garnet, staurolite, kyanite and sillimanite appear in the supracrustal belts. South of Mysore, considerable areas of charnockitic gneisses of granulite facies appear and supracrustal rocks are difficult to recognise. The isograds appear to transgress the tectonic grain, an arrangement which has been attributed to a northward plunge bringing lower levels to the surface in the south. At Sittampundi, in the Salem district of Madras, a pre-metamorphic stratiform anorthosite-gabbro mass is incorporated in the granulite-facies terrain. This body, resembling the earliest anorthosites of West Greenland (p. 55) is characterised by a very calcic plagioclase as well as by the occurrence of chromite-rich layers.

The Dharwar Series, which forms the supracrustal belts, includes a high proportion of volcanic rocks. A long succession, sub-divided into lower, middle and upper portions, has been recognised in Mysore by Rama Rao, Pichamuthu and others; but these divisions may not apply over large areas. In Mysore, lower Dharwar basic lavas and associated intrusions are followed by rhyolites and felsites. The middle division is a sedimentary group of conglomerates, pelites, carbonate-rocks and ferruginous quartzites. These are invaded by basic and ultrabasic intrusives and by granites. The upper Dharwars form a similar sedimentary assemblage.

The ferruginous quartzites are of the banded chert-magnetite type which, by surface weathering, give rise to rich hematite deposits. Associated with them are banded manganiferous metasediments which give manganese ores or manganiferous semipelites (*gondites*) carrying spessartite or rhodonite. Where the metamorphic grade is high, the Dharwar pelites display a characteristic mineral assemblage quartz-sillimanite-garnet-perthite (graphite) and produce rocks of the kind known in India as *khondalites*.

The migmatites, gneisses and granites which occupy the broad tracts between the supracrustal belts were evidently formed during more than one episode. Radiometric dates of over 3000 m.y. obtained from gneisses in the southern part of the massif suggest that some may represent the crystalline basement of the Dharwars. The majority of gneisses, including the widespread migmatitic *Peninsular gneiss* and the charnockitic varieties, have yielded ages of 2600–2300 m.y. which are thought to be related to the cycle of mobility involving the Dharwar Series. The *Closepet granite* – a complex massif measuring 300 km parallel to the tectonic grain and only 20 km across it, which crosses the regional zones of metamorphism and degrades charnockitic masses enclosed in it – has an apparent age of nearly 2400 m.y.

Gold mineralisation is associated, as in most Archaean provinces, with basic volcanics or intrusives in the supracrustal belts. The Kolar goldfield of Mysore is supplied by the Champion reef and other lodes situated near the contacts of schistose and more massive amphibolites and not far from outcrops of the Champion gneiss which has been regarded as a middle Dharwar intrusive.

Basic dyke swarms of the Dharwar massif. Swarms of basic dykes with a dominant north-north-west trend cross the metamorphic and granitic rocks of the Dharwar massif and are seen in an altered condition in the Eastern Ghats belt. Where emplaced in charnockitic country rocks they frequently have cloudy feldspars, a feature which has led to the suggestion that they were intruded before the closing stages of metamorphism.

2 Other ancient massifs

In north-west India, the *banded gneiss complex* of Rajasthan is a terrain of migmatites and granites, with remnants of supracrustal material, which appears

Table 7.1. THE GEOLOGICAL CYCLES OF THE SINGBHUM AREA

(based on Sarkar, Saha and Miller, 1969)

m.y.	Cycle		South of Copper belt thrust	North of Copper belt thrust
850	SINGBHUM	'Newer dolerites'	Granites, granophyres Gabbro-anorthosites, ultramafic intrusives	Granites, granophyres
1470				
			Kolhan Group (1600–1500 m.y.) mainly detrital sediments	Dalma lavas (basalts etc.)
				Singbhum Group (?2000 –1700 m.y.) (metasediments including banded iron formations, up to about 10 km)
			Dhanjori lavas (basalts) and quartzites (1700– 1600 m.y.) ·	
			←———— (unconformity) ————→	
2700	IRON ORE		Singbhum granite, minor basic intrusives (epidiorites)	
			Iron ore Group ⎰ Upper shales with sandstones and volcanics / Banded hematite jasper (B.H.M.) / Lower shales / Basic lavas / Sandstones and conglomerate	
			←———— (unconformity?) ————→	
3200			Granitic gneisses Older metamorphics with minor basic intrusives	

to underlie cover-successions of the Aravalli and Delhi belts (Table 7.2) and has been partially reworked during the evolution of these belts. Towards the east, scattered inliers of gneiss (Berach granite) link this complex with the massif of *Bundelkhand gneiss* in Uttar Pradesh (Fig. 7.1). Here, too, supracrustal remnants are enclosed in migmatites, gneisses and granites, the whole high-grade assemblage being traversed by prominent quartz-reefs and by dolerite dykes. Gneiss samples from these little-known terrains have yielded apparent ages of up to 2600 m.y. and the dykes of Bundelkhand have been dated at about 2500 m.y. We therefore regard the gneisses as remnants of a basement pre-dating the various later Precambrian mobile belts of the craton.

3 The iron ore belt, Singbhum

The ancient massif of Bihar and Orissa which lies to the west of Calcutta not only includes some of the oldest rocks of the craton but is also of exceptional

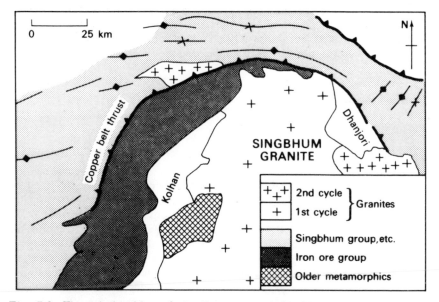

Fig. 7.2. The relationships of the Iron ore and Singbhum provinces in the north-east of the Indian craton (based on Sakar, Saha and Miller, 1969)

economic importance. The history of the massif reveals the effects of several geological cycles (Table 7.1). The iron ore belt itself has a north-north-east strike which is truncated both north and south by the margins of younger belts (Fig 7.2). The southern boundary is the edge of the Eastern Ghats belt. The northern boundary (the Copper belt thrust) appears to represent the front of a mid-Precambrian belt developed during what is known as the Singbhum cycle.

Exposures of the high-grade *Older Metamorphics* which include the oldest dated rocks of the Indian craton are partially enveloped by the Singbhum granite. The rocks of this ancient complex include pelites (quartz-mica schists), calc-magnesian schists carrying hornblende, basic igneous rocks (hornblende-schist) and granodioritic gneisses derived from the migmatisation of these types. K−Ar mineral ages for a number of rocks have ranged up to 3300 m.y. and Sarkar *et al.* (1968) assign a date of 3200 m.y. to the completion of the geological cycle in which they were formed.

The *Iron-ore Group* which is regarded as younger than the basement rocks, is a supracrustal sequence which, though everywhere folded, is locally almost unmetamorphosed. It forms two north-easterly belts flanking the Singbhum granite and its north-easterly folds are considered to be refolded in the vicinity of the Copper belt thrust. The principal components of the group are shales, sandstones and basic lavas, but there is in addition a formation of banded hematite-jasper which, when secondarily enriched, provides some of India's principal sources of high-grade iron ore. The *Singbhum granite*, whose age is given by Crawford as 2700 m.y., appears to be a late-tectonic batholith in which many successive units are outlined by concentric dome-like foliation patterns.

The massif of the Iron ore belt was, as far as can be seen, stabilised soon after the emplacement of the Singbhum granite. A northern part of this belt was reactivated in the Singbhum cycle (p. 133), while the southern part received a cover of basic lavas, followed by sandstones, conglomerates and shales of the Kolhan group, resting unconformably on folded rocks in two large basins (Table 7.1). Gabbro-anorthosite sheets are emplaced in the succession.

IV The Eastern Ghats Belt

1 The mobile belt

The *Eastern Ghats Belt*, following the eastern side of the Indian pensinsula, has remained relatively little-known since it is poorly exposed and economic ore deposits are few. The rocks of the belt show a fairly consistent NE—SW trend for over 1200 km. In the north, there appears to be a clearly defined front against the older iron ore belt. In the south a tectonic front against the Dharwar Massif is perhaps marked by belts of mylonite, or pseudotachylyte, traversing the ancient rocks of that massif. Between these regions, the marginal relationships of the Eastern Ghats belt are obscure, though in the Godavari Valley there are indications of a foreland with a flat-lying cover-series passing south-eastwards into folded rocks. In the extreme south, and more locally further north, the Eastern Ghats belt was itself apparently reworked during the 600—500 m.y. (Indian Ocean) orogenic period.

The metamorphic grade is high through most of the belt, and the predominant rocks are gneisses. Where metasediments have been recorded, they frequently include khondalites (p. 127). Charnockites are widespread and the type charnockite locality in Madras lies within the belt. There has been some doubt as to the dating of the type charnockites; Crawford, however, (1969) has obtained an isochron of almost 2600 m.y., suggesting an origin in cycles prior to that forming the Eastern Ghats. Both the Dharwar belt and the late Precambrian belt also contain charnockites and it would appear that charnockites were repeatedly formed, or at any rate rejuvenated, in southern India (Fig. 7.3). In spite of the uncertainties of dates and status — or, perhaps, because of them — this seems an appropriate place for a general note on the charnockite series.

2 The Charnockite Series

The term charnockite was applied by Holland (1900) to rocks of granitic composition with even-grained granular textures which characteristically contain hypersthene as the principal ferromagnesian mineral. The *charnockite series* ranges in composition from these acid types to intermediate, basic and ultrabasic rocks, covering much the same field as the calc-alkaline igneous rocks. All members of the series are characterised by hypersthene and share a number of other distinctive mineralogical features.

The type exposures form a number of small hills in the vicinity of Madras, but it was recognised by Holland that essentially similar rocks were widely distributed in southern India in areas that are now referred to the Dharwar

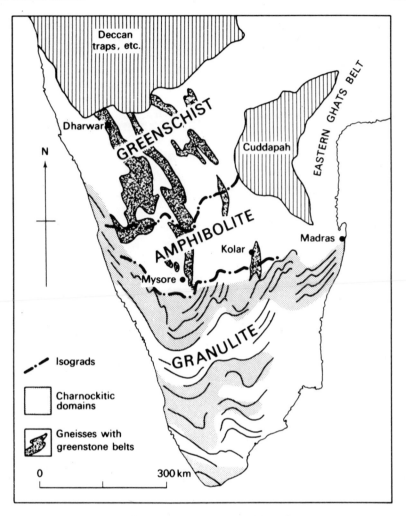

Fig. 7.3. Schematic map illustrating the relationships of metamorphic isograds and tectonic patterns in southern India (based on Pichamuthu)

and Eastern Ghats belts. These distinctive rocks were regarded by Holland as members of a consanguineous igneous assemblage. In the past sixty years, charnockitic assemblages have been described from every continent. The vast majority are found in provinces of high-grade regional metamorphism and although some appear to be intrusive, many are evidently metamorphic. The charnockite series proper is often associated with a more varied assemblage including metasediments such as the khondalites of the Dharwar and Eastern Ghats belts. The mineral assemblages of these associated rocks, and of the charnockites themselves, are those of the granulite metamorphic facies. Many charnockitic rocks are banded, foliated or migmatitic and the relationships of the more acid varieties with the more basic types, and with associated metasediments, frequently recall those between the granitic partner and its host-rocks in migmatitic complexes.

These features suggest that many charnockitic assemblages have originated under conditions of deep-seated plutonism in a dry crustal environment. Many metamorphic charnockites and allied rocks of granulite facies are now seen to belong to polycyclic complexes formed by the regeneration of older metamorphic rocks. Such an origin would be consistent with the distribution of many Indian charnockites, as well as with that of the allied rocks of the Mozambique belt of Africa (Part II) and the Grenville province of Canada (p. 87); it is of interest that few known charnockites have yielded isotopic ages of less than about 400 m.y.

3 Post-orogenic deposits

In the region north-west of Madras, an area of little-altered supracrustal rocks lies to the west of the Eastern Ghats belt (Fig. 7.3). The rocks of this basin — the *Cuddapah formation* followed unconformably by the *Kurnool formation* — rest unconformably on crystalline rocks of the Dharwar province. Lavas at the base of the Cuddapah give dates of 1370 m.y., and a diamondiferous kimberlitic intrusion in the formation a date of 1340 m.y. The succession is composed mainly of quartzitic sandstones and shales with minor conglomerates and limestones.

The Cuddapahs are intruded by basic sills and by quartz veins locally carrying lead–zinc–copper minerals. Towards their eastern margin they are folded and locally overridden by rocks of the Eastern Ghats belt and are invaded by small granites, syenites and pegmatites; they appear, therefore, to have been laid down before the termination of mobility in the Eastern Ghats belt and may be regarded as late-tectonic deposits.

V The Aravalli-Delhi Belt

The *Aravalli-Delhi belt* extends north-north-eastward through Rajasthan to Delhi for a distance of some 800 km (Fig. 7.1). In this belt, three major units related to three cycles are recognised and are given, with some other information, in Table 7.2.

The *Banded Gneiss Complex* underlying the Aravalli System has been reactivated at least twice since its initial formation more than 2700 m.y. ago. The Aravalli System is best developed in the southern part of the belt where an unconformable junction with the gneiss can be demonstrated. In includes thick basic volcanics near the base, associated with and followed by pelites with quartzites and limestones. Traced into Gujerat, the Aravalli merges into a series which includes iron- and manganese-bearing phyllitic members. A complex set of folds in the Aravalli defines an arcuate trend which diverges towards the south from the trend of the younger folds affecting the Delhi system. Widespread migmatisation associated with metamorphism locally makes the Aravalli difficult to distinguish from its reactivated basement. This metamorphism is tentatively dated at 2100–1900 m.y. by Crawford (1970).

Table 7.2. MAJOR UNITS IN THE ARAVALLI–DELHI BELT

m.y.	
? 1400–1100	(cratonic cover of Vindhyan sediments)
	←———————— *Unconformity* ————————→
? 1600–1000	DELHI orogeny, metamorphism and granitisation Delhi System: (metasediments)
	←———————— *Unconformity* ————————→
	Raialo Series: (metasediments)
	←———————— *Unconformity* ————————→
c. 1900	ARAVALLI orogeny, metamorphism and granitisation Aravalli System: (metasediments and volcanics)
	←———————— *Unconformity* ————————→
>2700	Banded Gneiss Complex

The *Raialo Series* and the *Delhi System* were deposited unconformably on folded and metamorphosed Aravalli rocks, or on the gneisses of the basement, and were themselves folded and metamorphosed during episodes preceding the emplacement of post-orogenic granites and pegmatites which yield ages ranging from 1650 to about 1000 m.y. Both the Raialo and Delhi successions, themselves separated by an unconformity, are largely of orthoquartzite facies, consisting of current-bedded quartzites and pelites, with conspicuous groups of marble and calc-silicate rocks.

The Delhi System occupies a rather narrow NNE belt, often no more than 50 km in width, which, though superimposed in a general way on the older Aravalli belt, is conspicuously discordant at its southern end. The unconformable contact with the older basement, marked by a deformed conglomerate, is preserved in some localities; but at the eastern margin of the Delhi belt a tectonic contact with the basement is marked by a broad zone of mylonites and cataclastic rocks. The folding of the Delhi System is complex and the metamorphism is of low to moderate grade. A few intrusions with the appearance of late-tectonic granites, such as the Erinpura and Malani granites, have given isotopic ages ranging from 1650–950 m.y. A few alkaline intrusions, including nepheline-syenites and carbonatites, have given dates ranging from 1500–1000 m.y.

VI The Satpura Belt and the Singbhum Cycle

The *'Satpura belt'*, emerging from beneath the Deccan Traps in the region of Nagpur, follows an east-north-east trend for some 800 km towards the Ganges delta.

The more westerly part of the Satpura belt includes two folded supracrustal formations – the *Sakoli Series* occupying an area in the south, and the *Sausar Group* forming an arcuate zone farther north. Both appear to rest unconformably on a complex of mica-schists, amphibolites and gneisses (the *Amgaon Series*) which forms a massif separating their main outcrops. Biotite and hornblende from this series have given dates between 1630 and 1430 m.y. which are regarded as those of the Amgaon orogeny, metamorphism and granitisation.

The *Sakoli Series* usually shows a relatively low degree of metamorphism, consisting mainly of chloritic, micaceous and epidotic schists, quartzites and occasional marbles. Rocks of the *Sausar belt* are of higher metamorphic grade and migmatites are widely developed, giving apparent ages which range down to less than 1000 m.y. The sedimentary facies is unusual in that fine-grained detrital and chemical sediments predominate throughout a thick succession: marbles, calc-silicate rocks and dolomites occur at several levels and manganese-rich sediments, sometimes associated with ferruginous quartzites, are prominent in at least one division. These metasediments, a valuable source of manganese-ore, show a banding comparable with that of chemically-deposited ferruginous quartzites and, like them, appear to be chemical precipitates. They carry metamorphic assemblages of Mn-bearing oxides or silicates; a common rock-variety in *gondite* characterised by spessartine garnet with quartz. Other common minerals are braunite, $3Mn_2O_3.MnSiO_3$, and the Mn-pyroxene rhodonite.

The eastern portion of the Satpura belt in Bihar is the zone affected by the Singbhum cycle, which crosses the northern end of the Iron ore belt (Table 7.1 and Fig. 7.2). The orogenic front against the ancient massif is marked by the northward-dipping Copper belt thrust which exhibits an important copper and uranium mineralisation. In the mobile belt to the north of the thrust, the cover-succession, possibly as much as 10 km in total thickness, is named the *Singbhum Group* by Sarkar and Saha (1963). It consists principally of pelites, orthoquartzites, turbidite greywackes, calc-silicate rocks and impersistent hematite-quartzites, and is followed by the *Dalma group* of basic, probably spilitic, lavas. These rocks, broadly equivalent to the little-disturbed Dhanjori and Kolhan groups which rest on the ancient massif south of the Copper belt thrust, are folded on arcuate trends parallel to the orogenic front and show medium or high grades of Barrovian metamorphism.

VII Cratonic Successions: The Vindhyan and other Formations

The Vindhyan of the northern part of peninsular India covers a vast area north of the Satpura belt and east of the Aravalli-Delhi belt, while rather similar formations occur to the west and south of these belts. The age of these formations has long been uncertain and some have been tentatively regarded as Palaeozoic. Isotopic dating, while establishing their Precambrian age, has shown that they do not form a natural group. In Jodhpur, 'Vindhyan' sediments rest on a thick series of rhyolites, possibly related to late-tectonic granites of the Delhi fold belt, which have yielded a date of 745 m.y. Further east, near the edge of the Bundelkhand massif, 'Upper Vindhyan' is intruded by kimberlite pipes with a minimum age of 1150 m.y. Clearly, the term Vindhyan has a facies significance rather than an age significance.

The bulk of the sediments consist of orthoquartzites and shales which are usually brown or red, sometimes showing conspicuous green reduction-spots. Many of the psammites are current-bedded or ripple-marked and some contain glauconite. These features, together with the presence of a few persistent limestones, suggest shallow-water marine deposition. The pattern of palaeo-currents in the eastern part of the outcrop suggests derivation from the south-east, that is, from the region of the Satpura belt. The Vindhyan is not, however, of typical molasse facies — conglomerates are not abundant and the psammites are seldom arkosic. It appears to represent platform-deposits laid down at various times after stabilisation of the main part of the Indian craton. Vindhyan sandstones are among the most quarried of Indian building stones and, in both ancient and modern times, have been used in the construction of some exceedingly ornate edifices.

8

The Australian Craton

I Preliminary

1 The structural uits of Australasia

The continental mass of Australia (Fig. 8.1) is made up of four great structural units, to which we may add a fifth represented in the fringing island arcs to the north and east. The oldest and most extensive unit is the *Precambrian craton* which comprises most of the central and western parts of the continent where Precambrian rocks are exposed beneath an intermittent Phanerozoic cover.

At its south-eastern margin, the second major structural unit, the *Adelaidean orogenic belt*, truncates the older Precambrian complexes. The Adelaide geosyncline received sediments of Infracambrian and Cambrian age and thinner sequences covering the same periods spread widely over the craton. The Adelaidean belt was subjected to folding and metamorphism in early Palaeozoic times, between 600 m.y. and 400 m.y.; it has much in common with early Palaeozoic belts of other Gondwanaland continents and is dealt with in Part II.

The third structural unit, the *Tasman orogenic belt*, occupies the eastern border of the Australian continent, from Tasmania to Queensland. This is a composite belt incorporating Palaeozoic and Mesozoic rocks of the Tasman geosyncline which were subjected to intermittent deformation and plutonism throughout the period from Cambrian to Cretaceous. Within the craton, broad sedimentary basins were developed at various times during the Phanerozoic, collectively providing the fourth structural unit of the continent. The *Great Artesian Basin*, with its prolongations northward into the *Carpentaria Basin* and southward into the *Murray Basin*, is the largest of these basins.

By the end of the Mesozoic, orogenic activity had almost come to an end throughout the Australian continent. Active mobile belts, however, remain in existence to the north and east, where the Circum-Pacific and Alpine-Himalayan orogenic systems converge. These still-active belts are represented by the *Indonesian islands* and associated oceanic deeps, and the archipelago of *New Zealand*.

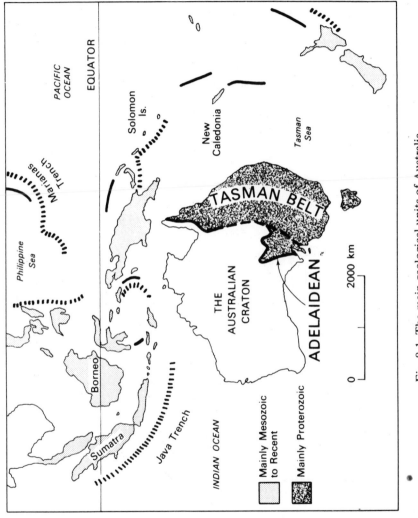

Fig. 8.1. The main geological units of Australia

2 The Precambrian craton

The Precambrian rocks which form the basement of the Australian craton are exposed in a number of large areas isolated from one another by outcrops of the Phanerozoic cover-series (Fig. 8.2). These isolated Precambrian areas, as well as certain structurally-defined provinces within them, are often referred to as 'blocks'. The presence of an extensive cover makes us prefer the term 'craton' rather than 'shield' for the stable mass as a whole. For so large a land-mass, the craton has rather a low relief, although some of the Precambrian blocks form mountain territory.

The Precambrian rocks of Australia are of special interest because they provide a record of Precambrian cratonic sedimentation spanning a period of at least 1500 m.y. The preservation of such cratonic sequences has led to suggestions that type successions, comparable with those which define the Phanerozoic stratigraphical column, might be established for the Proterozoic periods in Australia. Isotopic dating of volcanic groups and other key horizons has fixed the time-spans of certain successions and opened the way for regional

Fig. 8.2. Simplified map of Australia showing the main Precambrian units mentioned in Chapter 8 (indicated by capital lettering) and the main Phanerozoic basins of deposition

correlations. These developments will no doubt lead to the emergence of a new
stratigraphical nomenclature. In the meantime, it seems best to stick to a very
simple time-grouping into Archaean and Proterozoic events, with a dividing line
at 2400 to 2200 m.y. The more complex nomenclature adopted in the first
edition of the 1:6000 000 Tectonic Map of Australia (1962) involved recognition of Archaean, Lower Proterozoic and Upper Proterozoic, with further
distinctions between lower and upper portions of each Proterozoic division. This
system seems difficult to apply consistently at present and will not be followed
here. We shall, however, set aside the uppermost portion of the Precambrian (the
Infracambrian) to be described in connection with the Adelaide geosyncline in
Part II.

II Archaean Blocks

Archaean rocks form the greater part of two economically important blocks in
Western Australia, the *Yilgarn* and *Pilbara blocks* which are separated by the
east—west *Hamersley belt* of Proterozoic age (Fig. 8.2). Smaller areas of Archaean
or rocks which may be Archaean, crop out in the northern part of the state and
in the Northern Territories where they underlie Proterozoic rocks referred to in
later pages. In central Australia, the *Arunta* and *Musgrave blocks* which flank the
younger Amadeus trough are considered to include Archaean rocks, though they
have not so far yielded isotopic ages greater than about 1300 m.y.; and in South
Australia the *Gawler block* is also thought, though without confirmation from
isotopic work, to include Archaean rocks.

1 The Yilgarn and Pilbara blocks, Western Australia

The Yilgarn and Pilbara blocks will be treated together, since it is probable that
prior to the development of the intervening Hamersley belt they constituted a
single tectonic province; the Gawler block may represent the south-eastern
extension of the same province. The famous gold deposits of Kalgoorlie,
Coolgardie and Southern Cross — discovered in the last decade of the nineteenth
century and now almost worked out — provided an early stimulus for prospec-
ting which has been reinforced more recently by the discovery of substantial
reserves of nickel.

Two principal assemblages, resembling those which characterise granite-
greenstone belt provinces elsewhere, can be distinguished. A supracrustal
assemblage including a wide range of volcanics and associated intrusives, together
with subordinate sedimentary rocks, makes up a number of outcrops which
define a rather crude north-north-west trend. The areas which intervene
between these belts are occupied by granitic assemblages. The supracrustal rocks
of the greenstone belts, though strongly folded, are in some places barely
metamorphosed and retain many of their primary characters. Elsewhere,
especially in the 'Wheat Belt' towards the south-west of the Yilgarn block, rocks
apparently equivalent to those of the greenstone belts are seen to have been
metamorphosed in the granulite facies. A swarm of post-orogenic basic

intrusions dated at about 2400 m.y. provides a minimum date for the stabilisation of the province.

Greenstone belts. The record of volcanic activity provided by the rocks of the greenstone belts is of a type which has become familiar to us from the accounts of the Superior province of Canada and the Kalahari craton in southern Africa. It is necessary to emphasise that although the records in all three provinces are concerned with Archaean or Katarchaean events, they span altogether a period of no less than a 1000 m.y. and that greenstone belts were apparently formed intermittently through much of the early part of geological history. In Western Australia, volcanics of the Kalgoorlie region of the Yilgarn block are dated with some confidence at 2700–2600 m.y. (Compston and Arriens); volcanics in the Pilbara block are intruded by granites dated at about 3050 m.y.; both of these assemblages may prove to be younger than that of the Barberton greenstone belt of South Africa (p. 100).

The wide age-range of the greenstone belts complicates discussion of the relationships between the supracrustal rocks of these belts and the rocks of the intervening granitic areas. Small granitic bodies giving ages down to 2600 m.y. intrude even the younger greenstone belts. It has been widely assumed that the larger granitic tracts also represent intrusive bodies. Some of these tracts, however, were clearly in being before eruption of the younger volcanic sequences – for example, gneisses from the western part of the Yilgarn block give dates of up to 3100 m.y., whereas most of the volcanics of Kalgoorlie are not older than about 2700 m.y. A granitic basement of some kind is therefore present in certain areas, though there is as yet no firm evidence for the existence of such a basement prior to the formation of the oldest of the greenstone belts.

The components of the greenstone belts fall into two categories. One includes basic pillow-lavas, cherts, jaspilites, serpentines and layered basic-ultrabasic intrusive bodies. Some members of this association resemble the komatiites of the Barberton belt (p. 101) in being exceptionally magnesian. Nickel ores consisting of disseminated or massive sulphides, primarily pyrrhotite and pentlandite, are located in or alongside certain ultramafic bodies, for example at Kambalda. The second association is more varied, including acid lavas and near-surface intrusions, a range of intermediate and acid pyroclastics, polymict conglomerates, greywackes, shales, cherts and jaspilites. Groups in which these types are prominent (sometimes known by the old field name of 'whitestones') tend to be lenticular in contrast to the more persistent basic groups.

Where the stratigraphy is best-known, in the *Eastern Goldfields* of the Yilgarn block, there are indications that several volcanic cycles were involved in the build-up of a single greenstone belt. The succession recognised by Williams in the region east of Kalgoorlie (Table 8.1) records three such cycles in each of which basic and ultrabasic lavas and intrusives were followed by acid-intermediate volcanics associated with clastic sediments. In the Pilbara block, distinctions have been drawn between assemblages of basic volcanics (assigned to the Warrawoona succession) and sedimentary assemblages (the Mosquito Creek succession) which tend to rest unconformably on the local volcanics.

The thicknesses recorded in the greenstone belts are considerable. Williams gives a figure of over 8 km for the upper portion of Cycle 1 in the Kurnalpi

Table 8.1. THE ARCHAEAN SUPRACRUSTAL SUCCESSION IN THE KURNALPI AREA
OF THE EASTERN GOLDFIELDS, WESTERN AURSTRALIA

Cycle No.	Formation	Association	Thickness (metres)
3	Kalpini	basic volcanic	c.5000
	←————————— (unconformity) —————————→		
2	Gundockerta	acid volcanic-clastic (with turbidites and conglomerates	c.6500
	←————————— (local unconformity) —————————→		
	Mulgabbie	basic volcanic	c.4500
	←————————— (unconformity) —————————→		
1	Ginalbi	acid volcanic-clastic	>8000
	Morelands	basic volcanic	5500

(no known base)

Based on I. R. Williams, 1970 (Explanatory notes on the Kurnalpi 1:150 000 Geological Sheet, Western Australia Record 1970/71)

succession (Table 8.1) and an aggregate thickness of nearly 25 km has been claimed for the basic lavas with associated basic sills and sedimentary intercalations at Norseman, some distance further south, in the goldfields area. The meaning of these figures is arguable and we feel some doubt as to whether they should be taken to represent the vertical thicknesses of original lava-piles.

Structure. The mapping-out of key horizons and use of way-up criteria in the goldfields area has revealed a system of tight, but comparatively simple, folds of limited wavelength whose axial planes parallel the general north-north-west trend of the greenstone belts. These folds are sliced by dislocations and may in some localities be superimposed on isoclinal structures of larger wavelengths. Little is known of the structural relationships of the granitic regions. In the Pilbara block, granitic gneisses and granites appear to form broad dome-like bodies separated by tightly folded partitions of supracrustal rocks. In the goldfields area of the Yilgarn block, certain small granites occupy anticlinal cores in the greenstone sequences but others form discordant stocks rising high in the succession.

Mineralisation. The gold mineralisation in both Pilbara and Yilgarn blocks is concentrated in basic rocks of the slightly metamorphosed greenstone belts. The mineralised zones are structurally controlled; at Kalgoorlie, for example, gold-telluride and gold-quartz lodes are located in the dislocated limb of a small fold, the principal host-rock being an early sill known as the Golden Mile dolerite. Alteration associated with mineralisation is dated at about 2675 m.y., which is not far from the age of granites intrusive into the supracrustals. In the Pilbara block, pegmatites carrying tin, tantalum, beryllium and lithium are associated with granites which may date back to about 3000 m.y.

Post-tectonic basic suite. The Yilgarn block is crossed by a swarm of basic dykes which have a roughly east—west strike, perpendicular to the tectonic grain. Although most members of this swarm are dolerites of familiar types, there are also some large and unusual intrusions. The *Jimberlana* 'dyke' is in places more than 2 km in breadth and shows a near-horizontal layering of bronzititic, noritic, gabbroic and anorthositic varieties which define a number of basins along its length. The *Binneringie* 'dyke', over 2 km broad and some 300 km in length, is also strongly differentiated but exhibits a near-vertical layering. These remarkable bodies recall the Great Dyke of Rhodesia and Muskox intrusion of northern Canada. The date of about 2400 m.y. obtained from the Jimberlana intrusion shows that their intrusion came near the end of the Archaean cycle, soon after the stabilisation of the province.

2 The Musgrave and Arunta blocks

The inhospitable central portion of Australia reveals, beneath a patchy cover, two blocks made largely of high-grade metamorphic rocks and granites flanking the Amadeus basin which acted as a region of subsidence in late Proterozoic and Palaeozoic times. The character of the crystalline complexes encouraged the belief that they were of Archaean age, but this belief has not so far been confirmed radiometrically. The *Arunta block* to the north of the Amadeus trough undoubtedly has a complex history. Two metamorphic groups have been distinguished: an apparently older Arunta group of gneisses, basic rocks and high-grade metasediments sometimes of granulite facies, and a younger Riddock series mainly of amphibolite facies. Muscovite-rich pegmatites in this series supply the bulk of Australia's mica output.

The *Musgrave block* to the south is also largely made up of gneiss complexes including granitised greywackes with prominent kinzigitic marker-horizons, quartzites, calc-gneisses and early basic sills. Parts of the complex are charnockitic and certain later intrusions (for example, that of Ernabella) are also charnockitic in character. An important multiple and differentiated basic intrusion, the Giles layered lopolith showing a range from dunite to anorthosite, intrudes the gneiss complex.

3 The Gawler block

Much of the Gawler block (Fig. 8.2) is made up of gneisses, some migmatitic, some charnockitic, which envelop remnants of resistant supracrustals. In the southern part of the Eyre Peninsula, two groups tightly folded on north—south lines have been distinguished. The lower *Flinders Group,* with an estimated thickness of some 7000 m, consists of granitic gneisses and migmatised metasediments with marker-horizons of quartzite and dolomite. The succeeding *Hutchison Group,* some 2000 m thick, is made mostly of micaceous schists, with quartzites, dolomites, amphibolites and migmatites. Hematite-quartzites in the Middleback Ranges are enriched to form high-grade iron ores.

III Proterozoic Regions

There appears to be evidence of major plutonic activity over much of northern Australia in early Proterozoic times (dates for these events range from 2000–1750 m.y. in different localities) and this activity may prove, as knowledge advances, to constitute a key-episode in Proterozoic history. An early Proterozoic cratonic basin which received an important succession prior to these events is represented in the Hamersley belt of Western Australia. Plutonic events younger than the key-episodes are recorded in several blocks; some of these mid-Proterozoic episodes, ranging from about 1700 m.y. down to about 1000 m.y., were concentrated in clearly defined mobile belts, but others appear to have been sporadically developed and are of rather doubtful significance. Cratonic deposits ranging over the same mid-Proterozoic time-span are widely distributed in the northern and central parts of the continent. The topics to be dealt with can be grouped as follows:

1   Early Proterozoic mobile belts:
    The *Halls Creek belt, East Kimberleys,* Western Australia.
    The *West Kimberleys,* Western Australia.
    The *Pine Creek belt,* Darwin–Katherine area, Northern Territories.
    The *Willyama complex,* South Australia.
2   Early Proterozoic cratonic basin:
    The *Hamersley belt (Mount Bruce Supergroup)* Western Australia.
3   Later Proterozoic plutonic events and undated events:
    in northern Australia, *Mount Isa.*
    in south-western Australia, *margins of the Yilgarn block.*
4   Mid-Proterozoic cratonic successions:
    in northern Australia, the *Carpentarian* and rocks of the *Kimberley area.*
    in the *Hamersley belt, (the Bresnahan and Bangemall groups)*
    in central and southern Australia, the *Gawler block.*

1 Early Proterozoic mobile belts

In the Kimberley region of Western Australia, a squareish massif (the main portion of the *Kimberley block*) is flanked on two sides by early Proterozoic mobile belts which appear to be moulded around it (Fig. 8.2). Although the basement of the massif is entirely hidden by Proterozoic deposits, there is little doubt that it represents an Archaean block stabilised before the consolidation of the marginal mobile belts. These marginal belts — the *Halls Creek mobile belt* at the south-east side of the block and the belt forming the King Leopold Ranges in the *West Kimberleys* — were themselves largely stabilised in early Proterozoic times. Granites in the Halls Creek belt have yielded dates of up to 1970 m.y., suggesting that plutonic activity spanned the period around 2000 m.y. in which world plutonism was generally at a low ebb.

Supracrustal rocks of the Halls Creek Group include basic volcanics, sandstones, greywackes and subordinate dolomites in the East Kimberleys (where basic-ultrabasic intrusions occur) and a more restricted range of clastic sediments in the western belt. The dating of an intrusive pegmatite at 2700 m.y. suggests that these rocks were laid down in Archaean times; they have indeed some of the characters of Archaean greenstone belts and locally carry small deposits of gold. Folding took place on lines broadly parallel to the length of each belt; complex intersecting structures appear near their meeting-point and have been variously interpreted as indications of contemporaneity or of successive development. Metamorphism is of greenschist facies in many areas, though the grade rises to granulite facies in parts of the Halls Creek belt. The abundance of granite, some thought to be of anatectic origin, and the rather wide distribution of andalusite, suggests the operation of a high geothermal gradient.

In the Darwin–Katherine area, and around Tennant Creek and the Davenport Ranges, in the Northern Territory, remnants of the early Proterozoic Pine Creek mobile belt protrude through a later Proterozoic and Phanerozoic cover. The geosynclinal successions of this belt — the *Pine Creek sequence* or *Agicondian* in the region south of Darwin, and the *Warramunga group* of Tennant Creek — rest on an Archaean basement dated at >2400 m.y. at Rum Jungle. Granites emplaced during late-orogenic or post-orogenic events have given dates in the range 1820–1700 m.y. (lower figures are regarded as apparent ages by geochronologists, e.g. Arriens (1971)). The mobile belt includes highly mineralised areas carrying deposits with Au, Sn, Pb–Ag, W, Cu, and U.

The Agicondian of the Darwin–Katherine area reaches some 7 km in thickness and consists of coarse clastics of greywacke facies, merging laterally into shales, dolomites and cherts. Algal formations are particularly well-developed along the western boundary-fault of the mobile belt where they cover a ridge of supposed basement. The Warramunga group around Tennant Creek consists largely of greywackes and fine detrital sediments. In both parts of the belt, basic sills intrude the successions. Late greywackes and other psammites with occasional volcanic intercalations (*the Davenportian*) mark the final stages of infilling of the geosyncline. Faulting was active during the formation of these final deposits and, indeed, much of the Pine Creek basin is bounded and laced by faults. Folding of the cover is not particularly severe and metamorphism is of low grades. The basement outcrops in the northern part of the belt are circular in plan and could represent incipient mantled gneiss domes. The *Rum Jungle complex* of basement granites which yield Archaean ages shows little sign of regeneration. The uranium ores of Rum Jungle are located not far above the base of the Proterozoic cover, the general setting being reminiscent of that in which the uranium-deposits of the Witwatersrand System occur (p. 105).

In eastern South Australia, rocks east of the Adelaidean mobile belt are important from an economic point of view. These rocks, long ago called the

Willyama complex by Mawson (Willyama is the aboriginal name of Broken Hill), are interpreted as a geosynclinal sequence that was deformed, metamorphosed, migmatised and invaded by various granitic rocks prior to the deposition of Infracambrian glacigene sediments. Both these younger sediments (known by the local name of the Torrowangee Series) and the basement were modified during the Adelaidean orogeny.

The metamorphic grade in the basement seems highest in the Broken Hill area where Binns (1964) has recognised three zones trending SW—NE parallel to the regional structure. In the highest-grade zone, basic rocks appear as hornblende-two-pyroxene gneisses, pelites as sillimanite-orthoclase gneisses and granitic gneisses as orthoclase- rather than microcline-bearing varieties.

A variety of granitic rocks occurs in the Willyama complex. It seems likely that we are concerned with a series embracing migmatitic and syntectonic types at one extreme and massive post-tectonic types at the other. A conspicuous feature is the abundance and ubiquity of pegmatites, some of which reach a breadth of half a mile; Mawson held that 'in no other part of the world can pegmatite formations occur on a more extensive scale.'

We can now turn to the silver-lead-zinc mineralisation, which has made Broken Hill famous. The galena-blende ore is mixed with high-temperature minerals and the deposits have been regarded by some as replacements along a sheared and folded zone in highly-metamorphosed rocks. Others have argued for a syngenetic origin in which the ores were supplied by submarine volcanic sources and later subjected to deformation and metamorphism. Broadly speaking, it is possible to distinguish two types of Pb—Zn deposits, the more important *Broken Hill type* and the subsidiary *Thakaringa type*. From lead-isotope studies, a number of workers have suggested that the Thakaringa type resulted from partial regeneration of ores formed at about 1600 m.y.

2 An early Proterozoic cratonic basin: the Hamersley belt

The east—west Hamersley belt, some 500 km long and 300 km broad, separates the Yilgarn and Pilbara Archaean blocks in Western Australia and is occupied by the deposits of the so-called *Nullagine basin*. It provides superb outcrops of one of the world's most remarkable cratonic successions (Table 8.2). The lower members, together known as the *Mount Bruce Supergroup*, rest unconformably on the Archaean, but are themselves dated at 2200—2000 m.y. in age. The higher members, the *Bresnahan* and *Bangemall Groups*, are unconformable on folded rocks of the Supergroup and are dealt with on p. 151.

The *Mount Bruce Supergroup* forms an early Proterozoic sequence of at least 10 km which is preserved over much of its outcrop in an almost unmodified condition. Gentle folding is widespread, but severe distortion and disruption are confined to a few areas, as is metamorphic reconstitution. The outstanding feature of the succession is the occurrence of banded iron formations comparable in many ways with early Proterozoic iron formations of Labrador, Lake Superior and the Transvaal (Plate IV). The extraordinary character of these deposits – unmatched in the later stages of geological history – have been

Table 8.2. THE PROTEROZOIC SUCCESSION OF THE HAMERSLEY RANGES
(based on McLeod, 1966)

BANGEMALL GROUP		thickness variable but locally over 6000 m: mainly stromatolitic dolomites and sandstones, locally conglomerates, acid lavas: apparent ages of about 1100 m.y. from black shales and lavas	
		——— unconformity ———	
BRESNAHAN GROUP		thickness put at 13 000 m conglomerates followed by current-bedded sandstones.	THOLEIITE DYKES AND SILLS
		——— unconformity ———	
MOUNT BRUCE SUPER-GROUP	WYLOO GROUP	3000 m: unstable shelf-deposits, conglomerates, sandstones, siltstones, dolomites and basalts: contemporaneous acid intrusions dated at 2020 m.y.	GRANITE INTRUSIONS, FOLDING
	HAMERSLEY GROUP	2600 m: mainly chemical deposits, jaspilites, cherts, dolomites and banded iron formations: a rhyolite group near top dated at 2000 m.y.	
	FORTESCUE GROUP	2000–5000 m: mainly basalts and pyroclastics, with sandstones, cherts, dolomites: contemporaneous intrusions dated at c. 2200 m.y.	
		——— major unconformity ———	
ARCHAEAN BASEMENT OF PILBARA BLOCK			

revealed in admirable studies by members of the Geological Survey of Western Australia (e.g. McLeod, 1966, Trendall and Blockley, 1970).

The Fortescue Group which underlies the banded iron formations (Table 8.2) records clastic deposition and vulcanicity in an unstable basin showing considerable lateral variation in environment. The Wyloo Group which overlies them likewise provides evidence of instability and lateral variations. The intervening Hamersley Group on the other hand (Table 8.3) presents a picture of chemical deposition under astonishingly uniform conditions which prevailed over an area of not less than 140 000 sq km. Isopachs drawn on one division reveal outward thinning in every direction and Trendall envisages deposition in an almost enclosed basin. Cherts with varying amounts of magnetite, iron carbonates, hematite, stilpnomelane and riebeckite are the principal deposits, along with a dolomite, numerous thin 'shales' (rich in stilpnomelane and considered by Trendall to consist largely of fine volcanic debris) and a group of acid volcanics. The bedding of the cherts is extremely regular and in certain divisions individual layers only a few centimetres thick have been identified over distances of 100 km and more. A very fine lamination in many cherts is regarded as an annual layering, and consideration of this structure led Trendall to conclude that deposition took place at the rate of 30 cm of compacted chert per 2000 years. Apart from this lamination, a chert 'mesobanding' on a scale of a few centimetres, and a larger-scale layering defined by 'shale' partings, the structures

Plate IV. The Hamersley Range, western Australia, showing the outcrop of banded iron-formations of the Hamersley Group.

Table 8.3. THE HAMERSLEY GROUP:
an example of Proterozoic banded iron formations
(based on McLeod 1966, Trendall and Blockley, 1970)

Approx. thickness (metres)	Formation	Components
230	Boolgeeda iron formation	*BIF, chert, 'shale'
up to 630	Woongarra volcanics	rhyolite, dacite, pyroclastics
160–530	Weeli Wolli formation	BIF, 'shale'
500–730	†Brockman iron formation	BIF, chert, dolomite, 'shale'
17–100	Mount McRae Shale	'shale', siltstone, chert, dolomite
33	Mount Sylvia Formation	BIF, 'shale', dolomite, chert
260	Wittenoom Dolomite	
7–200	Marra Mamba Formation	BIF, chert, 'shale'

* BIF = Banded iron formation: total Fe averages over 20%. Hematite ores enriched during and after folding contain over 60%Fe and provide the principal iron ore reserves of Australia.

† Detailed study shows that sixteen 'shale' bands provide markers recognisable throughout outcrop of the formation: late-tectonic iron enrichment provides the hematite-rich ores of the Mount Newman and Mount Tom Price areas.

of the banded iron formation are almost entirely of diagenetic origin. The development of pod structures of characteristic form and arrangement is the principal result of post-depositional changes in the cherts. Details of the succession in the Hamersley Group are given in Table 8.3, as an example of the banded iron formations so characteristic of early Proterozoic times (see p. 187).

3 Later Proterozoic metamorphic complexes

Mount Isa. The Mount Isa area of Queensland is heavily mineralised and a variety of deposits carrying copper, silver–lead–zinc and uranium are emplaced in Proterozoic rocks. A metamorphic and granitic basement (formerly regarded as Archaean but possibly of early Proterozoic date) unconformably underlies the main supracrustal succession of the area, which appears to have accumulated from about 1750 m.y. onward. At least two younger periods of deformation are recorded, the first and more important being associated with granite-emplacement at about 1550 m.y.

The oldest members of the Proterozoic assemblage are the *Leichhardt Metamorphics*, formerly assigned to the basement. This group, consisting mainly of metamorphosed and migmatised acid volcanics, occupies a belt traversing the Mount Isa block from north to south which appears to have behaved as a divide, subject to repeated faulting and volcanic activity, during the subsequent development of the basin (Fig. 8.3).

The first basin which came into existence lay to the east of the divide. In it accumulated some 3000 m of acid lavas and pyroclastics thickening towards the west, interbedded with psammitic and pelitic sediments (the *Argylla formation*). The high-level Ewens granite, thought to be comagmatic with the volcanics, has given a reliable date of 1780 ± 20 m.y. which is close to that of the Norris granite intruding volcanics at the base of the Carpentaria succession (p. 152).

After the formation of these rocks, subsidence began on the western side of the median welt. Thick uniform sandstones – the *Mount Guide* and equivalent quartzites – were deposited over most of the mobile area to both east and west.

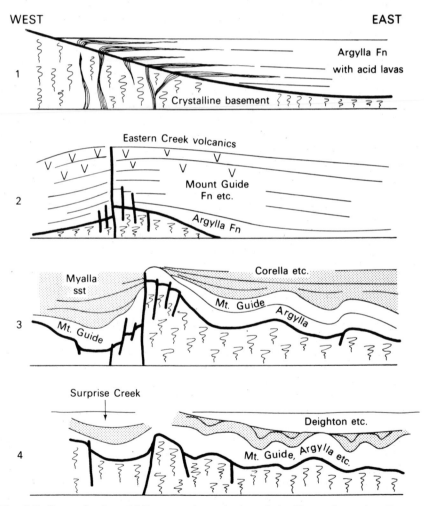

Fig. 8.3. Stages in the mid-Proterozoic evolution of the Mount Isa region (based on Carter and others, 1961)

Palaeocurrent observations suggest that the northern part of the median welt rose above sea level and for much of the subsequent period the two basins developed as separate entities. The *Eastern Creek volcanics* (basalts and shallow-water sediments) and their probable equivalents which follow the quartzites, appear to have been erupted along the divide and flowed down east and west into the adjacent basins.

Towards the close of the vulcanicity, widespread disturbances rucked up the floor and cover of the eastern basin into broad folds of north—south trend. The succeeding sequence — which can be called the Corella group after a prominent member — was laid down on an irregular surface; pelites fill the initial hollows and pass up into assemblages of carbonates, pelites and psammites, finally ending with extensive sandstones. In the western basin, fault-controlled subsidence allowed the accumulation of psammites such as the Myalla sandstone, which reaches a maximum of 6—7 km.

After deposition of the Corella sediments, the first period of orogenic deformation resulted in strong folding on north—south axes and uplift of rocks in the eastern basin with much transcurrent faulting. Deformation was accompanied by metamorphism of low-pressure type associated with migmatisation and the emplacement of granites. Disturbances in the western basin were much milder and are registered mainly by unconformities in the succession. Nearly 7 km of dolomites, sandstones and shales, forming the *Surprise Creek group*, were laid down in this basin after the onset of folding. As the median welt became less prominent, the *Deighton quartzites* were deposited to the east of it in a region of strong relief (Fig. 8.3).

Further east—west compression then ushered in the second period of orogenic deformation, resulting in complex folding, thrusting and transcurrent faulting. This phase was marked by regional metamorphism and by the intrusion of syntectonic and post-tectonic granites, often elongated in a north—south direction. Dating of both granites and metamorphic rocks has given ages of about 1450—1400 m.y.

The rich copper mineralisation of the Mount Isa block is mostly located near basic igneous rocks in fractures traversing pelitic rocks. The genesis of the Ag—Pb—Zn ores is controversial; a prominent view is that they are syngenetic, though epigenetic concentration may have taken place later.

Reworked margins of the Yilgarn block: the Esperance-Albany-Darling belt. Near Albany, towards the southern margin of the Yilgarn block, the general southerly trends of Archaean structures swing to east and west, parallel to the south coast. Traced eastward to Esperance, the trend becomes north-easterly and continues so for a considerable distance to the eastern margin of the block. This marked structural change is coupled with an increase in metamorphic grade, producing granulite facies assemblages and defines a unit of Proterozoic age apparently formed by regeneration of the Archaean complex. Infolded among reworked granulites and gneisses is a strip of high-grade pelitic metasediments interpreted by Wilson (1969) as the fill of a geosyncline of Proterozoic age.

The northern limit of reworking is made by deep-seated movement-zones styled the Fraser, Stirling and Brewer 'faults'; according to some workers, the charnockitic rocks of granulite facies are products of deep-seated plastic deformation and heat transfer in these movement-zones. A pegmatite cutting charnockites of the Fraser Ranges has been dated at about 1280 m.y., a date taken by Wilson and others (1960) to be that of the waning stage of a major movement on the Fraser 'fault'.

At the south-west corner of the Yilgarn block, these east—west structures make an orderly right-angled bend and continue northward for several hundred kilometres parallel to the large *Darling fault* which lies immediately to the west. The pelitic zone already mentioned turns northward in sympathy and is believed to be recognisable for 650 km to the north. Isotopic dates for metamorphic and igneous rocks along this western margin give ages of about 1100 m.y.

4 Mid-Proterozoic cratonic successions

Sequences laid down under moderately stable conditions after the culmination of the early Proterozoic orogenic cycle at about 1800 m.y. are preserved in three principal areas of north and west Australia — the Carpentaria province on the south-west side of the Gulf of Carpentaria, the Kimberley basin and the Hamersley belt — and on the Gawler block of South Australia. The rocks of the first-named locality have provided the material for some remarkable attempts to match stratigraphical and isotopic dating and are worth considering in a little detail. Those of the other localities will be dismissed more briefly first.

The broad Kimberley basin, as already noted, overlies an Archaean craton flanked by early Proterozoic mobile belts. The later Proterozoic cover is almost undisturbed where it rests on the craton but quite strongly faulted and folded near the margin of the Halls Creek and West Kimberley belts; movements on faults parallel to the tectonic grain appear to have continued in these belts right up to Palaeozoic times.

The lowest member of the succession, the impersistent *Whitewater volcanics* consisting mainly of acid lavas and pyroclastics, appears to be closely related to the later granitic bodies of the Halls Creek belt and has yielded an isochron age of 1820 m.y. It is followed by some 2—3000 m of sandstones with subordinate conglomerates, pelites and ironstones (including the ore-formation of Yampi Sound), together with an important group of basic lavas. These groups cover the greater part of the Kimberley block, providing a record of shallow-water deposition in fairly stable conditions. The history of intermittent deposition in the area is carried up to early Palaeozoic times by outliers of unconformable Adelaidean glacigene sediments, algal dolomites and sandstones, and by still younger (possibly Cambrian) plateau-basalts.

In the Hamersley belt the tectonic disturbances which followed the accumulation of the Mount Bruce Supergroup were in turn followed by the deposition of the younger part of the fill of the Nullagine basin. The *Bresnahan* and *Bangemall Groups* (see Table 8.2 for a summary of lithology) are quite severely folded and even metamorphosed along certain tracts, but are preserved elsewhere with little disturbance.

The crystalline rocks of the Gawler block in South Australia are overlain by unfossiliferous cover-sequences of uncertain age. One such is the *Moonabie Grit*, a clastic series with thin pebble-beds. It is followed, with slight angular unconformity, by the *Corunna Group* of conglomerates, shales and flagstones which, with the underlying rocks, are intruded by a group of feldspar-porphyries, the *Gawler Porphyries*, cropping out over an area of some 19 000 sq km. The Moonabie Grit has been tentatively correlated with the late Proterozoic Adelaidean, but doubt is thrown on this correlation by dates of 1500–1600 m.y. obtained from the overlying conglomeratic and volcanic group.

The Carpentaria province. In the region bordering the Gulf of Carpentaria, some 13 km of little-disturbed Proterozoic supracrustals rest on a crystalline basement represented by the rocks of the Darwin–Katherine area already described. The later phases of activity affecting the basement are securely dated at about 1800 m.y. Some of the overlying supracrustals and the basic sills which intrude them have been dated by McDougall and others (1965) and the results provide a remarkably complete documentation of the evolution of an early cratonic basin (Fig. 8.4).

The members of the *Carpentarian succession* are grouped by McDougall and his colleagues as follows:

(1) A basal series of acid volcanics appear to have been extruded soon after the initiation of new basins, in regions of disturbance. In the south-east of the province the comagmatic Norris granite gives a date of 1790 m.y. which is experimentally indistinguishable from those of the latest basement-granites. Hence it is suggested that vulcanism followed rapidly on uplift and erosion of the basement. In most localities, the volcanics are separated from the main sequence of sediments by an unconformity, but in the Darwin–Katherine area, volcanics and sediments interfinger.

(2) Overlying the volcanics is an assemblage of pelitic and carbonate sediments with intercalated volcanics. The main member of this assemblage, the *Tarwallah Group*, reaches some 6000 m thickness in the centre of the basin. Glauconitic beds high in the group have yielded a date of 1570 m.y.

(3) The succeeding division follows conformably on the Tarwallah group in the centre of the basin where it reaches a thickness of 5000 metres. On the flanks of the basin it is unconformable on the underlying rocks and has a thickness of only a few hundred metres. Its most important member is the *MacArthur Group* of dolomites, sandstones and cherts; a reef-complex shows stromatolites and sponge-spicules in abundance. Galena, syngenetic with a dolomite near the base, gives a model age of about 1560 m.y., while a glauconitic dolomite of an equivalent group in the Darwin–Katherine area gives an apparent age (regarded as a minimum figure) of 1255 m.y.

(4) *The Roper Group* which follows is largely confined to certain mobile troughs within the original basin. It rests on an eroded surface and consists mainly of shales, siltstones and sandstones reaching 2000 to 5000 m in thickness. Oolitic ironstones are of importance in at least one trough. Dated glauconitic sandstones some 500 m above the base of the group give mimimum ages of 1390 ± 20 m.y.

After deposition of the Roper Group and before further warping and faulting, tholeiitic dolerite sills averaging about 100 m in thickness were intruded in the northern part of the province. Somewhat altered specimens gave an acceptable date of 1280 m.y. for the time of intrusion. Deposition was later resumed in a

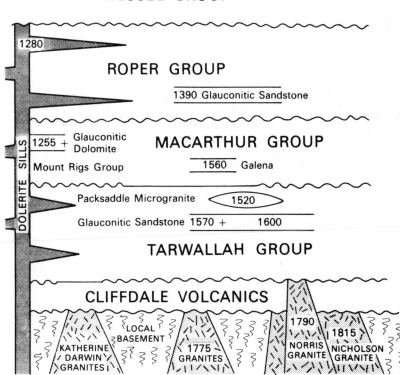

Fig. 8.4. A dated Proterozoic succession (the Carpentarian) in northern Australia showing the positions of dated marker-horizons (based on McDougall and others, 1965)

basin bordering the Arafura Sea where the Roper group is covered unconformably by the *Wessel group* of sediments, dated on glauconite at 790 m.y. but considered, from the presence of *Scolithus* borings, to be probably of Cambrian age.

During this account, it will have become obvious that some of the dates assigned to the Carpentaria groups may require further investigation; some geochronologists, indeed, regard dates obtained for glauconites from Precambrian rocks as suspect. Nevertheless, glauconite dates for the Carpentarian fit well with those obtained by other methods for associated volcanic and intrusive rocks. The broad consistency of the results may be due to the stability of the area in later times. The unusual thickness and completeness of the cratonic succession indicates that further work may lead to important advances in Proterozoic stratigraphy.

9

The Cratons of South America and Antarctica

I Preliminary

To deal with the early histories of the continents of South America and Antarctica in the one chapter is not a matter of pure expediency. It has long been recognised that the Andean fold-belt comes ashore in Grahamland (the Antarctic Peninsula) and, with the increase of exploration, many fundamental similarities in the constitution of the two regions have become manifest. Before embarking on our account of the Precambrian rocks of the two continents, we shall as usual give summaries of their geological histories, from which these similarities will become apparent. Geological knowledge of these difficult terrains is necessarily uneven so that our generalised descriptions must be provisional. For the same reason, we shall carry our descriptions on to include the latest Precambrian and earliest Palaeozoic rocks, rather than leaving these to be treated separately in Part II.

II The Geological make-up of South America

South America exhibits three contrasting topographic regions that are intimately controlled by their geological constitutions (Fig. 9.1). The coign facing the Atlantic is a region of plateaux, 600–1200 m above sea-level, separated into Guyanan and Brazilian portions by the Amazon Basin. This basin merges to the west with the second topographic element, an interrupted belt of broad plains, usually reaching only 2–300 m in height, that separates the eastern plateaux from the Andean mountain complex. This third element flanks the Caribbean on the north and thence runs for almost 7000 km along the Pacific border of the continent to Cape Horn; volcanic peaks in the Andes culminate in Ojos de Salado, over 7000 m high, in Chile. These three regions display structures of ages decreasing from east to west, principally Precambrian in the eastern plateaux,

Palaeozoic in the median plains and Mesozoic and Tertiary in the Andean mountain-complex.

In the eastern plateaux, Precambrian rocks are exposed in the Guyana and Brazilian shields. The oldest parts of these are metamorphic complexes that have been dated at over 2500 m.y. and have usually been called Archaean by South American geologists. In the Guyana shield, an extensive sedimentary formation

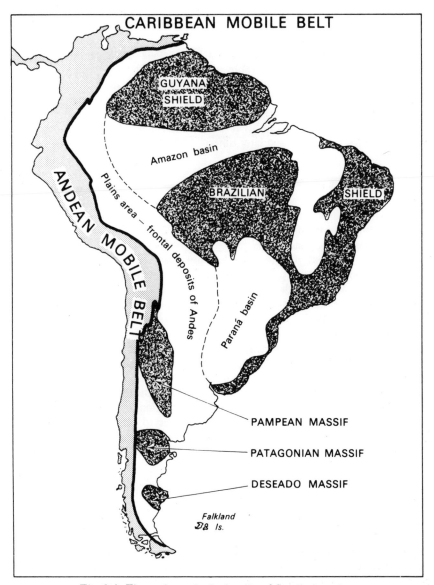

Fig. 9.1. The main geological units of South America

resting undeformed on the old complex has itself been dated at 1700–2000 m.y. In parts of the Brazilian shield a later Precambrian series has been affected by Precambrian orogenic movements. Especially along the Atlantic border from north-eastern Brazil to Uruguay, still younger orogenic events involving rejuvenation of older rocks between 600 and 400 m.y. are recorded. After the deformation and plutonism of Archaean and later Precambrian times, the entire shield was uplifted and warped into broad *intracratonic basins* (Fig. 9.1) in which thick sequences of Palaeozoic and Mesozoic sedimentary and volcanic rocks were accumulated. These piles have undergone no serious orogenic deformation.

In both the plains and the Andean belts, there are encountered many scattered occurrences of metamorphic rocks that have been variously regarded. Those in the Andes are difficult to interpret and some may be as young as Mesozoic. The massifs of the Pampean Range, Patagonia and Deseado (see Fig. 9.1), arranged in echelon across the junction of the two zones in Argentina, are confidently claimed as Precambrian blocks that have remained stable and emergent during Palaeozoic and later times.

III The Guyana Shield

The Guyana shield occupies the whole of the three Guianas — Guyana (formerly British Guiana), Surinam or Dutch Guiana, and French Guiana — and extends into the neighbouring states of Venezuela, Colombia and Brazil, thus involving a variety of nations and tongues that has not made for a coherent geological account. For the most part the old rocks form a high plateau, culminating in Roraima, some 3000 m high, and extending from the great bend of the Orinoco eastwards to the Atlantic; west of the shield, Precambrian rocks appear beneath Palaeozoic sediments in the deeper river-valleys of the Llanos plains of Colombia.

Classification of the constituents of the Guyana shield was for long based on their degree of metamorphism, high-grade metamorphism being taken to indicate an Archaean age, low-grade a Proterozoic. The advent of isotopic dating has shown that the old classifications require drastic modifications. For instance, the widespread non-metamorphic Roraima formation has been assigned by different workers to every period from Proterozoic to Cretaceous, with strong preference for the Triassic; isotopic dating has revealed that it is in fact at least as old as 1700 million years. It now constitutes an extremely useful time-marker, providing a lower age-limit for the diverse granitic and metamorphic complexes on which it rests (Fig. 9.2).

1 Assemblages older than the Roraima formation

There appears to be general agreement among South American geologists that the pre-Roraima rocks can be broadly grouped into two categories: *basement plutonic complexes* of high metamorphic grade and *sedimentary and volcanic successions* of lower grade (Table 9.1).

Table 9.1. PRINCIPAL PRECAMBRIAN GROUPS OF THE GUYANA SHIELD

Guyana (British Guiana)	French Guiana and Surinam	Venezuela
Basic intrusions in Roraima Formation 2000–1700 m.y.		
Roraima Formation	little-disturbed clastic sediments	
←——————————— *unconformity* ———————————→		
Bartica Assemblage (granitic migmatites locally gold deposits *c.*2000 m.y.)	Granites of Caribbean orogeny, 2200–1900 m y.	*Pastora Formation* folded about 1300 m.y.
Barama-Mazaruni Assemblage (supracrustal host rocks of Bartica assemblage)	*Orapu System* with tilloids granites of Guyana orogeny.	*Carichapo Formation* mainly volcanic, metamorphism about 2040 m.y.
	Paramaca System (volcanics and sediments)	
Rupununi Assemblage (Kanuku granulites, migmatites, South Savanna granite, *c.*2000 m.y.)	*Ile de Cayenne Series* (granulites, migmatites, granites), products of Hylean orogeny.	*Imataca Complex* (granulites, gneisses, granites, *c.*3000 m.y.)

The Basement complexes. Complexes which may be safely regarded as Archaean have been described from Guyana, French Guiana and the Venezuelan portion of the shield. In the southern parts of Guyana the oldest rocks of the shield are believed to be the *Rupununi Assemblage* and certain granites and gneisses as yet not directly correlated (McConnell and Williams, 1967). The term *assemblage* was introduced by Rodgers and McConnell (1959) for a large-scale association of groups of formations that is considered to be best for the objective mapping of such ancient terrains as the Guyana shield. The Rupununi Assemblage consists of rocks of granulite facies, hypersthene- and biotite-garnet granulites, charnockitic rocks, migmatites and biotite-granites. Age-determinations for the South Savanna granite suggest that it dates from about 2000 m.y. (Snelling and McConnell, 1969). In French Guiana, the *Ile de Cayenne Series* of migmatised and granitised metasediments of granulite facies is invaded by several phases of granites. Limestones and quartzites form resisters in the migmatitic gneisses. The structure is diverse, the dominant fold-trends being NE–SW or E–W. The orogeny involved was called the *Hylean* by Choubert (1956). In Venezuela, the northern part of the shield, the pre-Roraima rocks have recently been described by Kalliokoski (1965). The oldest metamorphic complex is the *Imataca* for which dates of about 3000 m y. have been reported. Quartz-feldspar-gneisses are associated with a variety of metasediments of granulite facies.

Iron and manganese chemical sediments form marker-horizons, characterised by gondites and related rocks rich in iron-garnets, -amphiboles and -pyroxenes or in manganese-rich garnets and rhodonite. Migmatisation is rather local and some of its products are obviously mobilised. Granites are present only in areas of lower metamorphic grade and may be considerably younger than the remainder of the complex. The complex is transected by powerful faults along which broad belts of flaser-gneiss and mylonite are developed.

Archaean supracrustal successions. Diverse supracrustal rocks of varying metamorphic grade have been separated from the basement complexes throughout the shield, and geochronological evidence suggests that many of these were involved in orogenic mobility terminating at about 2000 m.y. A multitude of series and systems have been named from which we select only those that appear to be most profitable.

In Guyana, an important cycle of sedimentation, orogeny and plutonism seems to be established. The *Barama-Mazaruni Assemblage* (McConnell and Williams 1970) of sedimentary and igneous rocks records the onset of tectonic disturbance. The Barama Group is composed largely of pelitic sediments, some manganiferous, with intercalations of pebbly, conglomeratic and pyroclastic beds. The Mazaruni Group, following evenly on the Barama, is of greywacke facies, with considerable accumulations of acid lavas and pyroclastics and basic intrusions — it is the 'Volcanic Series' of earlier workers; Cahen and McConnell (1969) record a provisional date of almost 2600 m.y. for a granite cutting volcanics assigned to it. Orogenic disturbances in which the rocks were folded isoclinally on E–W trends, metamorphosed to low grade, and invaded by syntectonic and post-tectonic granites seem to date from a later period. The granitised products make the *Bartica Assemblage* of gneissose granites, migmatites and resistant amphibolites, formerly regarded as basement complex. A minimum date for the final stages of granite-formation is considered to be 2000 m.y. The gold deposits of Guyana are located at the margins of the granites against 'greenstones' of the volcanic succession.

In Fench Guiana and Surinam, the equivalents of the succession just described have been assigned to two cycles by Choubert (1956). In the first of these, the *Paramaca System* was deposited unconformably on the basement. It consists of altered lavas, pyroclastics and related intrusions (mostly basic), with pelitic, psammitic and carbonate intercalations and minor iron formations. The metamorphic grade is mostly of epidote-amphibolite facies; the rocks are strongly folded on axes varying between NNW and NNE and are much migmatised and invaded by great bodies of calc-alkaline granite — the orogenic period being the *Gayana* in Choubert's nomenclature. The second cycle started with the deposition of pelites and outpouring of rhyolitic lavas, followed by the deposition of coarser clastic deposits, among which are conglomerates possibly of glacial origin — the whole being assigned to the *Orapu System* by Choubert. This system is folded and foliated on NW or NNW lines and penetrated by large granites dated at 2200 and 1900 m.y. — the *Caribbean orogeny* of Choubert.

In Venezuela, a number of metamorphosed and folded assemblages flank the central core of the Imataca basement already mentioned. To the south, the *Carichapo formation* consists of basic lavas, some pillowy, with minor sedimentary intercalations, including manganiferous and jaspery types, and basic intrusions. The formation is folded, metamorphosed and penetrated by acid intrusions and has reached a staurolite—almandine sub-facies of metamorphism; the Imataca basement suffered a second metamorphism during these events for which a date of 2040 m.y. has been given by Kalliokoski. Unconformably on the eroded Carichapo formation, there was deposited a series of slates, grits and cherts, with some pyroclastic beds and occasional rhyolite flows which is named the *Pastora formation* and compared to the Marazuni Group of Guyana by Kalliokoski. It in turn has been subjected to orogeny and plutonism, the date of which is suggested as 1300 m.y.

2 The Roraima formation

The flat-lying Roraima formation, buttressed by basic sills, makes high table-lands in Venezuela and the frontier regions of Guyana and Brazil (Fig. 9.2). The bordering scarps are precipitous and give rise to some spectacular waterfalls such as the Kaieteur Falls with an initial drop four times that of Niagara.

The Roraima succession of several thousand metres consists dominantly of sandstone, mostly pink or red in colour and in places current-bedded: a conglomeratic base is prominent in Brazil and conglomerates and shales are common throughout the pile. The clastic rocks contain much feldspathic material and detrital diamonds have been recorded. Microscopic forms, possibly of organic origin, have been recorded from Guyana. In the Guianas, rhyolites intervene between the sediments and the underlying basement.

The Roraima Formation is intruded by a great series of basic sheets, sills and dykes. Sheets of gabbro and orthonorite reach a thickness of hundreds of metres and their intrusion has resulted in doming and folding of the country rocks. When these great basic bodies come to be investigated in detail, it is certain that much light will be thrown on the mechanism of intrusion and the processes of differentiation and contamination. Thinner intrusions of dolerite (diabase) are found as dykes in the basement rocks far beyond the present outcrop of the Roraima Formation. Thin dykes of acid material, not all derived from the basic magmas, are also encountered.

Isotopic dates are available in some number for dolerites intruding the Roraima Formation and for muscovite and biotite from the associated hornfelses. From these, a minimum age for the Roraima Formation of 1800—2000 m.y. appears to be reasonable. The igneous rocks themselves have given a single date of this order, suggesting that they may belong to the same period of basic magmatism as the Bushveld complex of southern Africa; other dates of about 1700 m y. may reflect loss of radiogenic elements, though Snelling regards them as being close to the date of intrusion.

IV The Brazilian Shield

The Brazilian craton underlies the whole of Brazil and extends through Uruguay to the Plate and into the neighbouring states to the west (Fig. 9.1). Phanerozoic sediments of intracratonic basins cover large tracts of the old rocks, but the

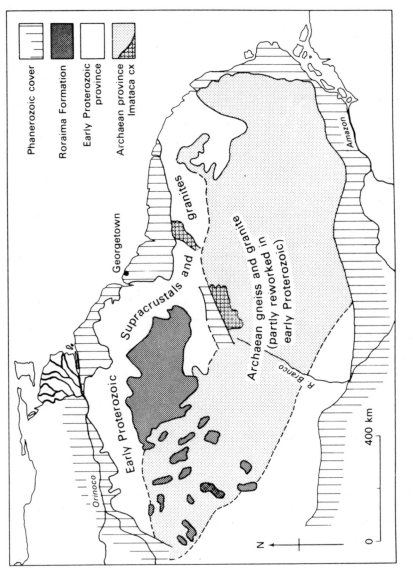

Phanerozoic cover

Roraima Formation

Early Proterozoic province

Archaean province Imataca cx

Orinoco

Georgetown

Supracrustals and granites

Early Proterozoic

Archaean gneiss and granite (partly reworked in early Proterozoic)

R. Branco

Amazon

400 km

N

0

Fig. 9.2. The Guyana shield (based on McConnell and Williams, 1970)

Brazilian shield made almost wholly of crystalline rocks forms the north-eastern salient of the country.

The Precambrian rocks of the shield were traditionally grouped by Brazilian geologists into 'Archaean' and 'Algonkian', a division admittedly based on metamorphic grade. As foretold by Oliveira in 1956 this grouping has been shown to be insecure. With the advent of even a minimum of age-determinations, many of the high-grade rocks thought to be Archaean have turned out to be of later age: further, reworking of old rocks appears to have taken place on a large scale, so that Archaean rocks have not always retained their early characters. In the present state of knowledge it seems most convenient to deal with the history of the Brazilian complex in three portions (Fig. 9.3). First, we consider very briefly those parts of the shield that are either proved to be Archaean or are presumed so to be. Secondly, we deal with later Precambrian events recorded in a great belt stretching from north-eastern Brazil south-westwards to the River Plate. The rocks concerned are mostly the *Sistema Mineiro* and its plutonised equivalents. Last, we glance at the *Lavras Series* which includes rocks of possible glacial facies and is probably of Infracambrian age.

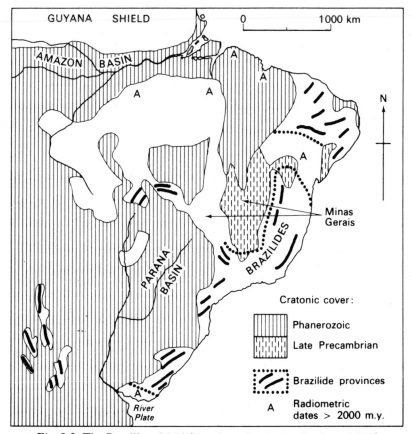

Fig. 9.3. The Brazilian shield (based on Cordani and others, 1968)

1 Archaean

Only a relatively small complex in the Brazilian shield has yielded radiometric age-data indicating Archaean granitisation and orogeny. During their invest-igation of the rocks of the Quadrilatero Ferriferro of central Minas Gerais, north of Rio de Janiero, Herz and co-workers (Herz, 1961) obtained dates of 2310–2510 m.y. on micas from granodiorites intruded into migmatitic gneisses in the centre of the Baçao complex. These Archaean gneisses are engulfed in younger granitic material, such as the Itabirito granite, giving ages of 1350 m.y.

2 Later Precambrian

The wide coastal belt of the shield includes most of the mineralised districts of economic importance and more information is available concerning succession, lithology and tectonics.

During later Precambrian times, successive geosynclines extending from north-eastern Brazil to Uruguay came into existence with general trends parallel to the present Atlantic coast. In these were deposited great thicknesses of varied supracrustal rocks which were subjected to plutonic and orogenic processes at intervals lasting into the beginning of Palaeozoic time. In these processes the Archaean basement must have been involved but, by reason of the extensive reworking of the old rocks and profound plutonic modification of much of their cover, the separation of basement and cover is not always possible.

Two major geological cycles at least have been proposed for the events in the coastal belt. These cycles have been classed by some as Middle Precambrian and Upper Precambrian but we prefer to use the more non-committal terms of *Pre-Minas* and *Minas and later* for them. The key horizon in this division (Table 9.2) is the great group of sedimentary rocks known as the Minas Series, so named because they contain vast ore deposits, especially of iron and manganese.

Pre-Minas events.   In Minas Gerais, Oliviera (1956) described the *Pre-Minas Series* as an assemblage of metasediments that were originally badly-sorted greywackes and grits, with thin carbonate, ironstone and tuff intercalations. This assemblage probably passes northward into better-sorted pelitic, psammitic and carbonate members with cherty and conglomeratic types and volcanics. Iron formations may locally be more than 1000 m thick and are sometimes associated with manganiferous sediments. The environment of the entire Pre-Minas Series is regarded as eugeosynclinal. The assemblage now shows strong westward thrusting of isoclinally folded packets and has been variably metamorphosed, mostly to greenschist facies; in places the rocks are of amphibolite facies, migmatitic granite-gneisses are developed and intrusive granites are present. Seven age-determinations on monazites from dykes cutting the Pre-Minas Series gave values of 1010–1100 m.y.

Minas and Later Events: the 'Brazilides'. The wide coastal zone shows a decrease in intensity of plutonic and orogenic activity inland. It is therefore fitting to describe first the succession in the unmodified regions of Minas Gerais where the rocks are well known (Table 9.2). These rocks, which rest unconformably on a variety of Pre-Minas assemblages, are the 'Algonkian' or later Precambrian of Brazilian geologists.

The *Minas Series* (Table 9.2) may have been deposited in a number of basins within a north-north-east tract more than 3000 km in length. The basal psammitic division rests, with local conglomerates, on an uneven erosion-surface and consists largely of current-bedded grits. These are followed by pelites of the Batatal formation and by the important Itabiri formation in a typical transgressive sequence. The chemical sediments of the Itabiri formation include the laminated ironstones known in Brazil as *itabirites* as well as dolomites, sparse manganiferous beds and fine detrital layers. The succession is concluded by the thick and variable Piracicata formation.

The Minas Series is strongly folded on north–south or north–east lines, the folds being overturned towards the west. Its metamorphism is mostly of greenschist facies but rises locally in the Piracicata formation where migmatitic gneisses appear. Intrusive granites give dates of 550–450 m.y. The *Itacolomi Series* which rests unconformably upon the folded Minas Series consists of current-bedded quartzites, with conglomerates containing pebbles of folded itabirite, succeeded by phyllites and an upper current-bedded quartzite. Among the psammitic rocks are the well-known flexible sandstones, *itacolomite*, on display in most geological museums. The series shows only a low grade of metamorphism and is broadly folded and penetrated by sporadic veins of aplite and porphyry.

We have already remarked that the amount of tectonic disturbance and plutonic modification in *the Brazilides* increases towards the Atlantic coast. Problems of correlation are raised by this lateral change; certain high-grade rocks (including charnockites) in the coastal tract thought by Ebert (1957) to represent transformed equivalents of the lower-grade later Precambrian rocks further west have yielded ages of about 2000 m.y. The Archaean floor is involved in reactivation, with widespread deformation and granitisation, and with the abundant development of charnockites and other rocks of granulite facies.

Table 9.2. SUCCESSION OF THE MINAS GERAIS, BRAZIL

Itacolomi Series	Quartzites, conglomerates, phyllites (450 m)
←——————————— *unconformity* ——————————→	
Minas Series	*Piracicata formation* quartzites, ferruginous quartzites, conglomerates, pelites, dolomites (3000 m) *Itabiri formation* laminated ironstones, dolomites, siltstones (<300 m) *Batatal formation* slates with minor cherts (<150 m) *Caraça Quartzite* (<200 m)
←——————————— *major unconformity* ——————————→	

Pre-Minas basement invaded by granites *c.*1350 m.y. and by dykes *c.* 1110 m.y.

The coastal zone of the Brazilides has been termed the 'internal zone' of the orogenic belt by Grabert (1962) in contrast to the less-disturbed 'external zone' to the west (Fig. 9.3). In the Serra de Mar and the Paraiba Valley, regarded as the centre of an orogenic belt of Alpine type, rocks of various primary ages are arranged in a series of large nappes trending north-west. Palingenetic mobilisation of basement rocks has taken place and the structural edifices formed have been invaded by abundant post-tectonic granites. A klippe of mylonitised charnockitic rocks is prominent in the Paraiba valley and further to the north-west, beyond Juiz de Fora, a high nappe of the same material is seen as infolds in packets of strongly-folded later Precambrian rocks displaced to the west on 'shingle' thrusts. Among the rocks of the eastern belt are charnockites and eclogites, together with kinzigitic gneisses containing garnet, sillimanite, cordierite and corundum.

Radiometric ages in the range 700–450 m.y. have been obtained from many localities in the entire tract extending from the north coast of Brazil almost as far as the River Plate. Isochron dates suggest that the main phases of activity took place at 650–600 m y., though K–Ar and Rb–Sr mineral dates, perhaps relating to post-tectonic events, are grouped around 500–450 m y. This coastal tract matches with that of the coast of West Africa which confronts it when allowance is made for the effects of continental drift and forms part of the extensive network of late Precmbrian to early Palaeozoic mobile belts discussed in Part II. Another branch of this system appears in the Pampean massif.

3 The Lavras series: Infracambrian

Non-fossiliferous, coarsely clastic formations that appear to be later than the post-Minas orogeny and earlier than Lower Palaeozoic sediments are present in many parts of the Brazilian craton. Correlation is always difficult with deposits of this type and opinion differs on the relative ages of many of the formations concerned. We take as representative the Lavras Series, which is widely developed in southern Bahia and northern Minas Gerais.

The Lavras Series in, for example, the Serra de Espinhaco, rests unconformably on the later Precambrian and consists of sandstone, conglomerates (sometimes diamond-bearing) and shales, with sporadic dolomitic limestones and layers of the carbonaceous rock shungite. The general opinion is that the series is of glacial facies and includes fluvioglacial gravels, tillites (e.g. *Sopa* formation) and related deposits that fill glaciated valleys in the post-Minas mountain ranges. Deposits of similar type have been recorded in a belt extending from central Bahia south-westwards for 1200 km to Sao Paulo and the Parana basin; the *Ribeira Series* of this latter region contains strongly-deformed tillites and shows extensive low-grade metamorphism. The glacial epoch of the Lavrás has been equated with that of Infracambrian age widely developed in other continents. In South America it is usually styled 'Latest Algonkian' or 'Eocambrian' though Stille allotted the Lavrás to the earliest Palaeozoic.

V The Antarctic Craton

As much as 99 per cent of the Antarctic continent is covered by ice and the scattered exposures of rock are notoriously difficult of access. But, as the continent occupies a key-position in the hypothesis of continental drift, any scrap of information about it may be precious.

A geographical division into East Antarctica and West Antarctica along the line connecting the embayments of the Ross and Weddel seas is customary and seems justified by many of the geological features discussed in this and later chapters. It has been estimated that the sub-glacial surface of East Antarctica would remain mostly above sea-level, adjusted to ice-melting, whereas West Antarctica would become an archipelago about an especially deep Byrd basin. This contrast reflects the geological status of the two units. East Antarctica has the characters of a continental craton, West Antarctica those of a young mobile belt. Crustal thicknesses are generally more than 35 km in East Antarctica and 30 km or less in West Antarctica.

1 The geological history of Antarctica

Whilst the sequence of geological events in the Antarctic continent can be stated with reasonable certainty, the actual distribution of the rocks recording these events can only be outlined provisionally. In Fig. 9.4 there is given some topographic and geological information that may help in our recital of the geological history (Adie, 1972).

The record of Antarctic geological history begins with the long sequence of events leading to the formation of the crystalline complexes which make the basement of the East Antarctic craton. From the nature of the terrain, the recognition of distinct tectonic provinces in this basement is difficult. The oldest radiometric dates so far obtained are in the region of 2000 m.y. and scores of age-determinations have shown a wide spread down to about 400 m.y. The partial overprinting of early dates by later events is suggested by this spread.

One age-province which can be set aside from the main body of the craton is that characterised by dates ranging from about 650 to 400 m.y., spanning the Precambrian—Palaeozoic boundary. These dates have been obtained from the *coastal regions* of East Antarctica, and from the *Transantarctic Mountains* near the western limit of the craton. In the coastal regions, they appear to be derived mainly from regenerated older rocks, but in the Transantarctic Mountains they are derived in part from supracrustal rocks of a cover-succession — the Ross Supergroup — laid down in a late Precambrian and earliest Palaeozoic geosyncline. It thus appears that the Precambrian craton was almost encircled by a mobile belt equivalent to that of the Brazilides (p. 163).

After the Lower Palaeozoic orogeny, the folded complexes of East Antarctica were elevated and eroded to produce a widespread surface, the *Kukri Peneplain*, on which were deposited the continental sediments of the *Beacon Group*, ranging in age from at least Lower Devonian to Jurassic. These sediments were intruded by a suite of Jurassic basic dykes and sills, the *Ferrar dolerites*. At some time later than the Jurassic, the basement and its cover of Beacon Group and Ferrar

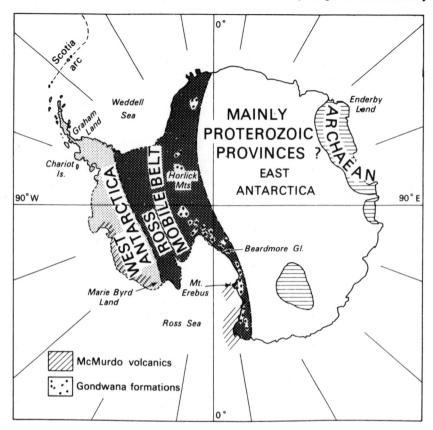

Fig. 9.4. The main geological units of Antarctica

dolerites was subjected to vigorous block-faulting and some of the uplifted blocks have been stripped, wholly or partly, of their cover.

The main part of *West Antarctica* is formed by a fold-belt *of late Cretaceous–early Tertiary age* that has marked similarity with the Andean fold-belt of South America, and is linked to it by the Scotia arc. As in the Andes, the fold-belt is based on a folded complex of Palaeozoic or Precambrian age. The main orogeny of late Cretaceous–early Tertiary age affects thick piles of Jurassic and Cretaceous sediments and volcanics and is characterised by calc-alkaline igneous rocks identical with the Andean intrusive suite of South America.

2 The older rocks of the Antarctic craton

Scattered information about the older parts of the East Antarctic craton comes from the reconnaissance studies of nunataks and ice-free inland areas, many of which have been carried out under conditions which were far from ideal. The records suggest that gneiss complexes of granulite facies are extensive. These complexes include some well-known charnockitic assemblages typified by those of Enderby Land, which gave its name to the plagioclase-rich charnokitic species of *enderbites*. The high-grade assemblages of Enderby Land are assigned by geologists of the Leningrad Research Institute for the Geology of the Arctic to a 'pre-Riphean' age-group partially reactivated in early Palaeozoic times. They include charnockitic pyroxene-gneisses and migmatites whose hosts are sillimanite-bearing quartzites and pelites and calcareous gneisses.

In the Miller Range of the central Transantarctic Mountains a less profoundly altered assemblage is represented by the Nimrod Group whose metamorphism is tentatively dated at about 1000 m.y. The group includes shallow-water sediments and basic volcanics, repeatedly folded and metamorphosed to amphibolite facies. Granites of the Ross cycle intrude the Nimrod Group which was modified during the early Palaeozoic activity.

The coastal regions of East Antarctica. As we have already seen, isotopic dating in these regions suggests that many rocks are polycyclic and permits the recognition of Precambrian and of early Palaeozoic events, though the division is not easy to apply in practice. In our account we can accept as our main guide the summaries given by Warren (1965, pp. 286, 293) for the exposures in the Australian sector between 89°E and 160°E. This area includes the U.S.S.R. station, Mirnyy (93°E), Wilkes Land, Terre Adelie (French) and George V Land (150°E), and has been investigated in considerable detail.

From the field-relations and the isotopic dates available. it appears that the Precambrian complex between 89°E and 160°E was derived from typical geosynclinal sediments, basic lavas and intrusions associated with them. Rocks of granulite facies were characteristically produced, typified by charnockitic gneisses, together with kinzigites, amphibolitic gneisses and amphibolites. Migmatites are widely but sporadically developed and age-determinations of their quartzofeldspathic components range from 1200 m.y. to 940 m.y. Other rocks of the complex have given ages up to 1540 m.y. In the Bunger Hills (95°E) and in other localities, a later period of migmatisation has been proposed with dates of 760 m.y. to 670 m.y.; with this migmatisation is associated the intrusion of basic charnockites. It should be remembered that Terre Adelie was the site of Stillwell's (1918) classic researches in which he defined the processes of metamorphic differentiation and metamorphic diffusion that have since been shown to play important roles in plutonic transformations.

Further along the coast, to the limit of the Australian sector, rocks similar to those just described are exposed in the Vestfold Hills (78°E), the Mawson area (63°E) and Enderby Land (45°–55°E). In the first locality, rocks of granulite facies give ages in the range 1104 m.y. to 1482 m.y. and, in the same region, a rock described as an arkosic sandstone gave two determinations of 1790 m.y., the greatest age so far obtained for an Antarctic rock. The 'Mawson Granite' of Mawson Land is a gneissic charnockitic granite which contains hypersthene-bearing xenoliths of finer grain; isolated masses of enstatite-cordierite-sapphirine rock have been described by Segnit (1957).

In the Norwegian sector, a variety of gneisses and migmatites is exposed along the coast of Dronning Maud Land (20°W–45°E). In the range Yamato Sanmyaku some 300 km inland, Kizaki (1965) finds an older complex of charnockitic rocks that have been subjected to later granitisation; basic dykes cut the complex and their amphibolite-facies metamorphism is attributed to the intrusion of later microcline-granites. Further west along the coast as, for instance, in the Sør Rondane Mountains (22–28°E), granitic gneisses, amphibole-rich gneisses and migmatites are encountered. Abundant marbles and calc-silicate rocks persist as resisters. Whilst the prevalent metamorphism is of granulite grade, rocks of lower grades are developed in areas of intense cataclastic deformation.

3 Infracambrian and early Palaeozoic events

We have just seen that many rocks of the East Antarctic complex have given dates of around 1000 m.y., or even older, and all opinion is tolerably certain that most of the complex originated in Precambrian time. But, nevertheless, early Palaeozoic dates of between 540 and 400 m.y. have been obtained for a variety of granitic gneisses, charnockites, migmatites and metasedimentary schists. In many localities, such rocks occur in association with intrusive granitic rocks which yield the same early Palaeozoic dates. It seems, therefore, that we are dealing with a zone subjected to extensive regeneration in early Palaeozoic times.

4 The Ross geosyncline and the Ross orogeny

In the Ross Sea area of the Trans-Antarctic Mountains, a thick series of metasediments, mostly metamorphosed to low grades, is folded on north–south axes and invaded by large syntectonic and late-tectonic granites of many kinds. The sediments form the *Ross Supergroup* (Warren, 1965) and somewhat similar rocks are known at wide intervals in the immense region east of the Trans-Antarctic Peninsula (see Fig. 9.4). It is a matter for discussion whether this extensive series (if it is a unit) was deposited in a single broad basin or in a number of more or less parallel troughs. In the type area of Victoria Land the sediments are thick and diversified enough to suggest the existence of a *Ross geosyncline*. The folding, metamorphism and granite-invasion that they have undergone is the *Ross orogeny* (Gunn and Warren, 1962) and the whole sequence of events from sedimentation to final uplift constitutes the *Ross cycle*.

The age-range of the Ross Supergroup is not securely established. Limestones containing *Achaeocyathus* and conglomerates, shales and acid volcanics believed to be of Cambrian or early Ordovician age are included in the Supergroup according to Grindley and Warren (1964). However, certain granites intrusive into metasediments of the supergroup have yielded ages of over 600 m.y., suggesting that deposition was interrupted by metamorphism and granite-emplacement prior to deposition of the early Palaeozoic succession. Younger granites give ages of 500–450 m.y. and are themselves probably early Palaeozoic.

In the Ross sector of the Trans-Antarctic Mountains, where the Ross Supergroup is best-known, the primary sedimentary types are greywackes, pelites and carbonate-rocks. The Teall greywacke of the McMurdo Sound area, for example, is at least 2000 m in thickness and is associated with some 3000 m of well-bedded impure limestones, black calcareous shales and fine greywackes – the Anthill Limestone of Gunn and Warren (1962). In the Wisconsin Range of the Horlick Mountains, greywackes, phyllites and impure quartzites are followed by a massive unit of acid rocks, interpreted as a metamorphosed tuff or ash-flow (the Wyatt formation) which has yielded a Rb–Sr date of 633 ± 13 m.y.

The Ross rocks are strongly folded on axes usually directed north–south and, for the most part, are of greenschist facies. In the McMurdo Sound area and probably elsewhere, however, much higher-grade rocks have original sedimentary characters like those of the Ross Supergroup and are regarded as equivalent to them. For example, the Koettlich Marble, presumably the extension of the Anthill limestone, consists of high-grade chondrodite-marbles and sillimanite-schists; this increase of metamorphic grade appears to be accompanied by an increase in the amount of granitic material present and its degree of intimacy with its country rocks.

The granitic intrusions into the Ross Supergroup and its possible equivalents have been grouped as the Granite Harbour or Admiralty complex. Most radiometric determinations suggest an Ordovician age, but rapakivi granites of the Wisconsin Range batholith have yielded Rb–Sr ages of 627 m.y., suggesting the occurence of an Infracambrian phase of granitic emplacement. The great Larsen batholith, over 350 km long, of the McMurdo Sound area is considered to be syntectonic; it possesses a weak foliation and encloses innumerable fragments of its country rocks – its average composition is granodioritic. The numerous post-tectonic granites and granodiorites of the same area are massive rocks devoid of foliation and form small plutons, some cutting the Larson batholith. Small, plug-like bodies of diorite penetrate the plutons, and dykes of micro-granite, microdiorite and lamprophyre round off the assemblage.

10

Problems of the Precambrian Record

I Purpose

Now that the early geological history of individual continental masses has been described, the way is open for a more general consideration of the events recorded during this immense span of time. Many of these events, of course, are of kinds which have recurred throughout geological history. Others, though also recurrent, seem to have been a little different at each repetition. Still others took place once for-all and therefore may be held to call in question the validity of the doctrine of uniformitarianism.

The immensity of Cryptozoic time relative to Phanerozoic time makes the present chapter an appropriate place to discuss some recurrent events, even though their final repetitions have still to be described. In a general way, we shall first consider happenings which seem to be cyclic and then pass on to those which seem to have been confined to the earlier chapters of earth history. As a preliminary to these discussions it will be useful to assemble some kind of synopsis of events in the several regions concerned. The upper time-limit of this survey generally falls at about 800 m.y., but varies slightly according to the local geology and the state of geological knowledge.

II Precambrian Geological Cycles

Figure 10.1 summarises the leading events described in Chapters 2 to 9: we know of no source for the corresponding information concerning the Pre-cambrian ocean basins. The chart is not intended to suggest direct correlations of more than a very few events, and even these landmarks are fixed only within wide time-limits. All that is implied in most columns is a relative time-sequence with some provisional date-control. The recognition of certain broad similarities between the sequences recorded in different columns and the identification of a few events which appear to have been common to several continents is as much as can be hoped for at present.

In dealing with the Precambrian record, we have constantly made use of the concept of geological or orogenic cycles. It is desirable at this point to enquire how far the cycles conform to any general pattern. From this enquiry, some indications of a turning-point round about 2800–2400 m.y. emerge. As far back as this period, there are records of geological cycles similar in many ways to the Phanerozoic cycles dealt with in Part II. The Svecofennide belt of the Baltic shield and the Coronation and Labrador belts of Canada, for example, record phases of initial sedimentation and vulcanism overlapping with periods of tectonism and plutonism which are terminated by episodes of granite formation and eventually by stabilisation. These belts came into existence at, or before, about 2200 m.y. and date from the early stages of the Svecofennide chelogenic cycle.

The records of Archaean and Katarchaean events are more ambiguous. It is not easy to fit the histories of such typical Archaean complexes as the Superior province of the Canadian shield or the Kalahari craton of southern Africa into a conventional cyclic pattern. The history of surface events is dominated by volcanic activity. The evolution of individual greenstone belts shows a characteristic sequence which has been repeated over different spans in different localities, but the scale and distribution of the belts are so different from those of later mobile belts that we hesitate to compare the greenstone belt cycle with the orogenic cycle.

III The Role of the Precambrian Cratons

A second point which is worth recalling is the evidence for the co-existence of mobile and stable crustal domains during cycles extending as far back as 2800 m.y. The broad contemporaneity of events recorded from adjacent mobile and stable domains was only established when isotopic dating techniques came to be employed; two examples cited in previous chapters are provided by the accumulation of the Witwatersrand succession (p. 105) over the time-span occupied by successive phases of mobility in the Rhodesian craton and Limpopo belt in southern Africa, and the accumulation of the middle Proterozoic cratonic succession of the Carpentaria province on northern Australia, while in the Mount Isa district a sequence of broadly the same age was being repeatedly metamorphosed, migmatised and invaded by granites. In each of these examples, the contrasts of sedimentary facies, of volcanic activity, style and extent of tectonic disturbance and of metamorphic activity were comparable with those which might be observed between stable and mobile domains of Phanerozoic age.

Although fundamental contrasts in the mobility of the crust were therefore already in existence in Archaean times (doubtless reflecting then, as now, a pattern of behaviour in the mantle) there are strong reasons for supposing that the distribution of mobile and stable tracts has changed radically in successive cycles. Such changes are evident even within the span of Phanerozoic times. The continental cratons of the Mesozoic and Tertiary eras are of continental

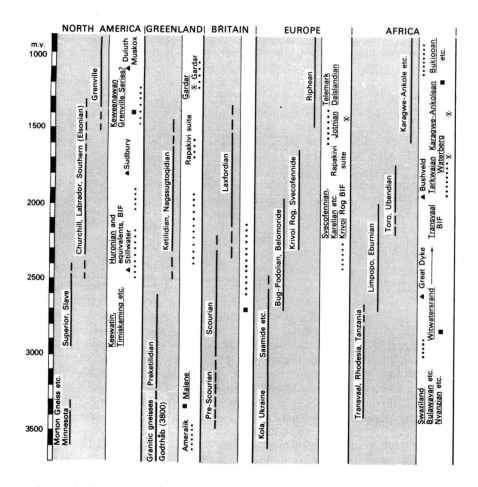

Fig. 10.1. Summary chart showing important geological events in each continent

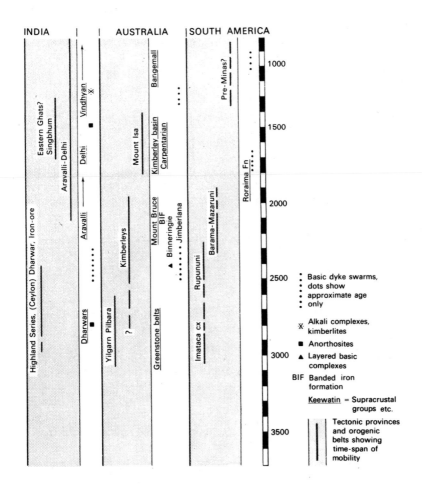

dimensions. The early Palaeozoic and late Precambrian cratons were very much smaller and were enmeshed in a network of mobile belts. Comparable variations in the dimensions of Proterozoic cratons can be recognised The Archaean cratons enmeshed in the early Proterozoic mobile belts were small; the Superior province of Canada (Fig. 4.2) provides a typical example The cratons which developed after the stabilisation of early Proterozoic mobile belts such as the Hudsonian belt of Canada at about 1800—1700 m.y. were very much larger, though perhaps not equal in size to the largest stable masses of the present day. From evidence such as this, Sutton (1963) has inferred a rhythmic variation in the extent of the stable cratons (Table 10.1). Many small cratons enclosed in a network of mobile belts characterised the early stages of the Svecofennide and Grenville chelogenic cycles, but the progressive stabilisation of mobile belts led, during each cycle, to the welding of small cratons into larger stable masses and the restriction of mobility to a few branches of the original network of belts. The extension of cratonic conditions is linked with other geological happenings considered in the next section.

Superimposed on this rhythmic fluctuation in the extent of crustal mobility, which according to Sutton has a periodicity of about 1000 m.y., a still longer-term, apparently unidirectional change can be perceived. Virtually all the surviving evidence for the first 1000 m.y. of recorded geological history is of the kind which would, in more recent geological contexts, be taken to indicate

Table 10.1. DEVELOPMENT OF CRATONS IN EARLY GEOLOGICAL TIMES

Chelogenic cycle	m.y.	
SVECOFENNIDE	1500	Progressive stabilisation of early Proterozoic mobile belts welds early Proterozoic cratons into larger stable masses.
	2000	Evolution of network of early Proterozoic mobile belts reactivates parts of stable areas and defines many small cratonic blocks, for example Superior and Slave province, Pre-Ketilidian massif, Archaean massif of Baltic shield.
SHAMVAIAN		Basic intrusions including regional dyke swarms in many cratonic areas.
	2500	Stabilisation of many Archaean provinces leads to development of first large cratonic blocks, represented in every continent.
	3000	First cratonic cover-succession begins to accumulate about 2800 m.y. (Witwatersrand succession of South Africa)
	3500	No known Katarchaean or early Archaean cratons.

crustal mobility. The Witwatersrand succession of the Transvaal, whose oldest components are tentatively dated at 2800 m.y., is the oldest supracrustal sequence which has survived almost undisturbed and unmetamorphosed and may be taken as an indicator of the earliest cratonic block so far identified: a block probably not much more than 1000 km in diameter. On the other hand, there is abundant evidence of the existence of smallish cratonic blocks at about 2500 m.y. The Great Dyke of Rhodesia (p. 113), the Jimberlana norite and its associates of Western Australia (p. 142) which cut through the folded and metamorphosed rocks of Archaean provinces and which are themselves unmodified, are dated at 2500–2400 m.y., while many unmodified dyke swarms and supracrustal sequences date back at least to 2300 m.y. There is a good case for suggesting that the characteristic pattern of cratonic blocks bordered by mobile belts, which has been a feature of the continental crust through most of recorded geological history, came into existence over the period 2800–2400 m.y., as a result of the first widespread phase of stabilisation of previously mobile tracts.

Prior to this period, it seems reasonable to suppose that the behaviour of the crust was governed by a different tectonic regime under which all continental crustal regions exhibited at least some of the types of activity which we now associate with mobile belts. The patterns which can most often be recognised in the surviving Archaean or Katarchaean provinces are typified by the association of greenstone belts with small massifs or domes of granitic or migmatitic rocks (see Figs. 6.3 and 6.7). There is some evidence of contrasts in the reactions of these components, sustained over long periods of time (see for example, the discussion of the Rhodesian massif p. 112). The granitic regions exhibited a tendency to rise. while the greenstone belt successions were pinched into deep synclinal screens between them. Temperatures sufficiently high to lead to the mobilisation of granite were attained in or at the margins of the granitic regions and steep thermal gradients allowed the survival of almost unmetamorphosed rocks in the interiors of the greenstone belts. Although some authors regard the granitic massifs as embryonic cratons ('protocratons' of Goodwin), their behaviour was more nearly akin to that of mobile granitic massifs of younger times, such as the mantled gneiss domes of the Svecofennide belt. Talbot (1968) interprets the pattern of granitic batholiths and greenstone belts in the Archaean massif of Rhodesia in terms of the effect of systems of small convection cells in the mantle beneath a crust some 35 km in thickness. Whatever view we take of these suggestions, it seems certain that crustal mobility in Katarchaean and early Archaean times took place under tectonic regimes which were fundamentally different from those of later geological times.

IV The Basic Interludes of the Cratons

The last general point concerns two Precambrian events which are possibly of world-wide significance — we may refer to them as the *basic igneous interludes* of the cratons. Intrusion and extrusion of basic magma has, of course been a recurrent feature since the earliest geological times. What distinguishes the two Precambrian interludes referred to here is the exceptional extent of activity

Table 10.2. BASIC IGNEOUS ROCKS OF THE CRATONS, c 2300 m.y.

Baltic Shield	Greenland NW Scotland	North America	Southern Africa	India	Australia	South America
Basic dyke systems in basement complex below Karelian cover.	Kuanitic, Kangamiut and other dyke swarms cutting Pre-Ketilidian massif. deformed in Ketilidian and Nagssugtoqidian belts: Scourie dykes cutting Scourian complex, deformed in Laxfordian belt. Scourie dykes give apparent ages of c.2200 m.y.	Dyke swarms cutting Archaean complexes of Slave province and Ungava peninsula deformed in Churchill province: Nipissing diabase in Huronian cover (2150 m.y.) Stillwater complex (?2450 m.y.) and dyke swarm in Wyoming.	Great Dyke of Rhodesia (2500 m.y.) Ventersdorp lavas (2300 m.y.) Bushveld igneous complex (2050 m.y.)	Basic dyke swarm in Dharwar province	Jimberlana and Binneringie 'dykes' and associated swarms cutting Archaean of Yilgarn block (2400 m.y.)	Basic dykes and sills in Roraima Formation of Guyana shield (c. 2000 m.y.)

recorded in the cratons. Some relevant details are assembled in Tables 10.2 and 10.3 which show that regional dyke swarms and major intrusions, associated in some instances with plateau-basalts, are represented in several widely separated continents.

Some of the rocks formed during these interludes have been well known for many years. It was not, however apparent that so many of them were formed within two comparatively short time intervals, until the results of radiometric dating both of the rocks themselves and of the orogenic cycles which preceded and followed their emplacement became known. The dates obtained spread over a time-span of several hundred million years – the earlier interlude falling roughly at 2300 ± 200 m.y., the later at 1200 ± 200 m.y. – and we infer that this spread was a real one. The igneous suites mentioned in Tables 10.2 and 10.3 can therefore only be regarded as coeval in the broadest possible way: their time-spans are short when set against the span of 1000 m.y. which characterises the long-term chelogenic cycles of crustal activity.

If the basic suites listed in the tables are to be thought of as products of connected igneous events, then the primal cause must have been one which operated on a global scale. This inference is strengthened by the fact that the basic interludes appear to coincide with periods during which orogenic activity was restricted as a result of widespread stabilisation and extension of the cratons (see p. 174). A connection with the terminal stages of the long-term Shamvaian and Svecofennide chelogenic cycles of Sutton (or with the Superior and Hudsonian regimes of Dearnley) may ,be indicated; the time of the basic interludes coincides with 'lows' in the histograms of isotopic ages of granitic and metamorphic rocks formed in mobile belts (Fig. 1.4).

Table 10.3. BASIC IGNEOUS ROCKS OF THE CRATONS, c.1200 m.y.

Baltic shield	Greenland	North America	Africa	Australia
Post-Jotnian diabases (1400–1300 m.y.), basalts of Karelia	Basalts of Gardar formation and Gardar dyke swarm (1400–1200 m.y.)	Coppermine basalts. Muskox intrusion, NW Territories Mackenzie dyke swarm, Canadian shield. Keewenawan basalts (1100 m.y.) Duluth gabbro (1100 m.y.)	Bukoban and associated volcanics and dyke swarms of E. Africa (1200–950 m.y.)	Basic sills in Carpentarian (c.1300 m y.

With these coincidences of timing in mind, it is interesting to look forward to the corresponding stages of the succeeding Grenville chelogenic cycle for comparison. As is shown in Part II, the last 200–300 m.y. have been marked by progressive extension of cratonic conditions in the continents and restriction of orogenic activity to comparatively few belts. Over the same period, basic magmatism has played an exceptionally important part in the stable regions, giving huge lava-plateaux such as the Deccan Traps of India and dyke- or sill-complexes such as those of the Karroo of Africa. Furthermore, much of this magmatism appears to have been related both spatially and in time to the disruption of the supercontinents and the concomitant opening up of the Atlantic and Indian Oceans.

In the most recent chelogenic cycle, therefore, abnormal basic magmatism in the cratons appears to have been connected with active phases of continental drift. While we need not necessarily assume that continental disruption accompanied the two Precambrian basic interludes, we can infer that they coincided with major periods of extension and that they mark, in a different form, the terminal stages of the earlier cycles.

V Sub-divisions of the Precambrian Record

The difficulties of making practicable divisions of the Cryptozoic eon will have become obvious from the foregoing chapters. Our device of using the geological cycle as the principal time-unit, building appropriate repetitions into larger units (the chelogenic cycles or megacycles) and obtaining smaller divisions from local evidence, is not a particularly tidy one. Many geologists would prefer to adopt time-divisions based on the stratigraphical record rather than on the records of crustal mobility on which we have relied.

A *Precambrian stratigraphical column*, if it is ever to be established, will probably be based on cratonic successions. The requirements are little disturbed and unmetamorphosed sequences, containing volcanic or other horizons suitable for isotopic dating. Two successions which go some way towards meeting these requirements are the Witwatersrand sequence of the Transvaal (2800–2050 m.y.) and the Carpentarian of northern Australia (1800–1400 m.y), which are summarised in Table 6.3 and Fig. 8.4 respectively.

The Australian Precambrian includes other sequences which may have a value as stratigraphical standards and the subject has been extensively discussed among Australian geologists. A general scheme is set out by Crook (1966) in a paper bearing the aspiring title 'Principles of Pre-Cambrian Stratigraphy'. Isotopic dates play the part of fossils in allowing correlation. Stratigraphical reference-points are chosen which mark the beginning of particular time-rock units in a carefully selected continuous succession. The time span of each time-rock unit extends from one reference-point to the next and is fixed within certain limits by isotopic dating of suitable horizons. Reference points in the succession are related where possible to geological events of regional significance, such as the initiation of a geological cycle or of a basin-filling. Crook considers that the application of such a method may eventually provide a scheme of Precambrian systems and periods capable of world-wide recognition, the imperfections of isotopic dating being no greater than those of dating by means of fossils in Phanerozoic periods. The restrictions imposed by the requirements of the method – especially the fact that it is, as yet, not applicable to successions in which even a low degree of metamorphism or disturbance has modified the apparent isotopic ages of the units – suggest that the proposed divisions will have only a limited value.

Two terms for broader divisions of Precambrian time, which still carry the overtones of long discussion, have been used a good deal in the foregoing chapters (Fig. 1.2). To these terms, *Archaean* and *Proterozoic*, we may add the less commonly used term *Katarchaean*. In the revised (1960) 2nd Edition of the *Stratigraphic Classification and Terminology* produced by the National Committee of Geologists of the USSR, the following criteria are laid down for the division of the Precambrian into Archaean and Proterozoic Groups (pp. 57–8, English summary). Most important are stratigraphical and structural unconformities of regional character, and especially significant are the truncation or deflection of older fold-belts *by younger. The presence or absence of pre-orogenic ophiolites, or of granitic rocks comtemporaneous with or younger than the orogenies is regarded as being of value, but grade of metamorphism is of doubtful significance. Isotopic age-data will be critical when they achieve their maximum reliability. It seems to us that these criteria are best satisfied, and the long-term rhythm of geological cycles is best expressed, by allowing the divisions between the Katarchaean, Archaean and Proterozoic portions of Precambrian time to fall at the ends of the appropriate megacycles. In particular it seems best to use the 2300 m.y. basic interlude as marking the close of the Archaean. This procedure conforms roughly to that followed in several other countries. The classification adopted by the Geological Survey of Canada, shown on p.56, agrees in using the geological megacycles as the basis for its time-divisions, but breaks the cycles at a different point: at the peak of plutonic activity in mobile belts, which is relatively easy to determine by isotopic dating.

VI Precambrian Tectonic Patterns and the Question of Precambrian Plate Tectonics

Displacements of continental masses relative to one another and collisions between converging masses were among the most spectacular events of the Mesozoic and Tertiary eras. The question arises as to whether these displacements for which the term *plate tectonics* is commonly used were related to a unique event or whether similar movements had taken place at earlier times. This can be examined in two ways – by considering whether reasonable patterns of orogenic belts, cratons and other major crustal features formed during the Superior and Hudsonian regimes are obtained when the displaced continental masses are restored to the positions which they occupied prior to the Mesozoic break-up and by palaeomagnetic investigations.

The approximate trend-lines in tectonic provinces formed during the Superior regime, when plotted on continental reconstructions such as those published by Bullard *et al.* (1965), or by Smith, Briden and Drewry (1972) reveal a smoothly flowing structural pattern. Dearnley (1966) identified a centre of convergence towards the present Arctic Ocean and a gentle symmetrical convergence in eastern Asia and concluded (p. 29) that 'It would seem that a number of periods of continental fragmentation are unlikely, since otherwise it would be a most unexpected and remarkable coincidence to find that the Precambrian fold-belts . . . still fitted a relatively simple pattern.'

Structural patterns of various types which were established in early Proterozoic times also appear to show a continuity over areas of continental, or even supercontinental, dimensions. Dyke swarms emplaced during the 2300 m.y. basic interlude (Fig. 10.2) seem to line up on a great circle crossing from northern Canada to Greenland and Scotland (Payne *et al.*, 1965). Transcurrent movement-zones of north-easterly trend traverse North America and may extend to

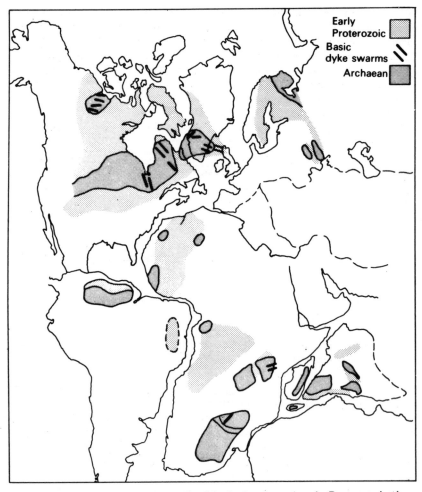

Fig. 10.2. Tectonic provinces stabilised in Archaean and early Proterozoic times, with some basic dyke swarms emplaced in the Archaean provinces before *c.* 1950 m.y. (plotted on a continental reconstruction of Briden and Smith, allowing for post-Palaeozoic continental drift)

Greenland and Scotland (Fig. 4.6). The major occurrences of banded iron formations (p. 187) fall in a sinuous belt which Goodwin (1973) considers to be continuous across present continental boundaries.

These features suggest that by early Proterozoic times coherent masses of continental crust had attained dimensions at least as great as those of the present continents. The preliminary findings of palaeomagnetic studies indicate that these large crustal masses were in motion relative to the magnetic pole, but suggest that they moved for long periods as units and did not undergo fragmentation and dispersal. The paths of *polar wandering* – which indicate changes in position relative to the pole – suggest that North America, Africa.

and probably other continental masses, changed direction several times during the Proterozoic era (Fig. 10.3), and may have collided with neighbours at these times.

There is, therefore, strong evidence that large continental plates were moving relative to the poles, and perhaps also to each other, from late Archaean times (*c.*2500 m.y.) onward. This does not necessarily indicate that the style of plate tectonics characteristic of the late Phanerozoic eras was already in operation. The complex network of mobile belts developed during the early stages of the Hudsonian regime in North America and Greenland, for example, appears to be situated very largely within a continental plate which palaeomagnetic data show to have been moving in broad terms as a unit, whereas late Phanerozoic mobile belts were sited at the junctions of plates moving independently. This contrast suggests that the large continental plates of early Proterozoic times lacked the strength and stability of present-day continental plates and suffered heating and deformation along mobile tracts developed entirely within their confines.

Ensialic mobile belts formed in continental crust far from plate boundaries have no real analogues today and such Proterozoic belts as the Nagssugtoqidian of Greenland, the Laxfordian of Scotland, and the western Churchill province of Canada may turn out to differ in many ways from the conventional model of a mobile belt.

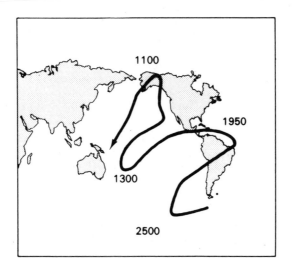

Fig. 10.3. Precambrian polar-wandering curve for North America (based on Donaldson and others, 1973). The figures give (in m.y.) the inferred ages of the 'hairpins' which record changes in the nature of movement of the continent

VII Reactivation and Continental Growth

From every continental mass described in the foregoing chapters there have been cited examples of the reactivation – or rejuvenation to use Eskola's term – of crystalline basements during later cycles of orogeny. At depth in the crust under conditions of high temperature and pressure, the basement has become

tectonically and chemically active. Such a rejuvenated basement is capable of
assuming new shapes and new structural patterns; in addition, it may be
chemically modified, receiving influxes of new material from below and
expelling emanations and intrusions into its cover. The modifications achieved
by these changes may be so great that the components of the basement become,
to all intents and purposes, new rocks. Such reconstitution at depth is in many
ways complementary to the surface processes of erosion and sedimentation.

The reactivated basements provide the favoured environment for the
development of certain rock-groups. The charnockitic suite with associated
rocks of granulite facies, appears on a regional scale most often within polycyclic
complexes such as those of the Mozambique belt of Africa (Part II) and the
Eastern Ghats belt of India. Certain large anorthosite bodies are considered to
have been generated or mobilised by reactivation (see pp. 85—7) and especially
Berrangé, 1965, for discussion). There is increasing evidence that many terrains
of banded gneisses, whether of granulite or of amphibolite facies, have been
repeatedly reworked. The relatively low water-content of a crystalline basement
at the start of regeneration may be expected to favour the development of 'dry'
mineral assemblages. We may recall the evidence given in Chapter 6 that
reworked Archaean complexes form large parts of the Proterozoic Churchill and
Grenville tectonic provinces of the Canadian shield. Late Proterozoic (Grenville)
basement rocks which may already have been polycyclic are incorporated in the
Palaeozoic Appalachian belt of North America. Such examples lead us to conclude that reworked continental crustal material has, from late Archaean times,
been the principal contributor to orogenic mobile belts. The evidence seems to
us incompatible with hypotheses which invoke the growth of continents
through geological time by addition of stabilised mobile belts formed largely on
oceanic basements.

One final topic which may be mentioned in this section concerns the orogenic
belts, which appear to be composed almost entirely of polycyclic crustal
material. The Grenville belt of North America and the Mozambique belt of
eastern Africa are structures of this type. There is no evidence that supracrustal
successions of 'geosynclinal' thickness accumulated on the site of these belts
during the early stages of the cycles of mobility in which their tectonic and
metamorphic patterns were established. The principal supracrustal formations
preserved in them have in some instances been identified as deposits laid down in
earlier cycles — for example, the banded iron formations incorporated in the
Grenville belt are assigned to the Hudsonian cycle (pp. 85—6).

The absence of cover-successions in such belts may of course, be due simply
to the effects of deep erosion in laying bare the basement. Such an
interpretation, in our view, would have to rely heavily on an assumption that
'geosynclinal' sedimentation and vulcanicity have been essential elements in the
evolution of mobile belts. We have seen (p. 181) that ensialic mobile belts with
distinctive geological relationships were developed in Proterozoic times. In these
belts, tectonic and thermal activity was not necessarily associated with the
accumulation of cover-successions of distinctive thickness or facies. Belts of this
type form a large part of the late Precambrian—early Palaeozoic network of
Gondwanaland (p. 99 and Part II). Most are characterised by polycyclic
assemblages of high metamorphic grade, often including charnockites or other
rocks of granulite facies.

VIII Evolution of the Atmosphere and Hydrosphere

Two groups of hypotheses concerning the history of the earth's atmosphere and hydrosphere have been advanced. The first regards them as derivatives of *primary* materials remaining from the primitive fluids that enveloped the molten earth, the other as *secondary* materials supplied by leakage or *outgassing* from the deep interior. Most authorities appear to favour the second alternative. concluding that any primary atmosphere was lost to the earth at a very early stage. The present atmosphere is notably depleted not only in the light elements hydrogen and helium but also in the heavy inert gases krypton and xenon, as compared with the solar system as a whole. These chemical peculiarities seem to indicate an almost total loss of any initial atmosphere and for our purposes we need go no further back than the stage at which the earth's home-produced atmosphere and hydrosphere began to accumulate.

The sources of this secondary material are the rocks of the crust, mantle and core. Rubey, in an influential discussion of this topic (1951) showed that the 'excess volatiles', over and above those likely to have been released by weathering, are principally H_2O, CO_2, Cl_2, N_2 and S. There is a marked similarity between these excess volatiles and the juvenile gases emitted by volcanoes, fumaroles and hot springs, or incorporated in igneous rocks. Rubey pointed out that the crystallisation of a 40-kilometre granitic shell could supply all the water in the oceans. He regarded the evolution of atmosphere and hydrosphere as continuing geological processes and invoked outgassing as the prime cause.

From the occurrence of water-laid detrital sediments in the oldest well-preserved supracrustal groups such as the Onverwacht of the Barberton greenstone belt and the Isua supracrustals (pp. 53, 56), it is clear that the accumulation of an atmosphere and hydrosphere had begun before the dawn of recorded geological history. The early atmosphere is thought to have been *anoxygenic*, to use a term coined by Rutten. Urey (1952) favoured a strongly reducing mixture of water vapour, hydrogen, methane, ammonia and hydrogen sulphide but later writers such as Cloud (1971) envisage an atmosphere dominated by H_2O, CO_2, CO, N_2, HCl, H_2 and S, which could be supplied directly from magmatic sources. Evidence of the reducing character of the atmosphere is thought to be provided by the occurrence in early Precambrian sediments of detrital grains of easily oxidised minerals such as pyrite and uraninite. The conversion of such an anoxygenic atmosphere into the *oxygenic* atmosphere of the present earth was bound up with processes linked to the evolution of life, which are considered in the next section. There is considerable disagreement among the experts as to the length of the period that elapsed before the oxygen content of the atmosphere approached its present level. Berkner and Marshall suggested in 1965 that this stage was delayed until about 600 m.y., when it ushered in the explosive evolutionary phase indicated by the Phanerozoic fossil record. Other authorities favour a much earlier date for the build-up of oxygen and envisage the existence of an oxygenic atmosphere at, or well before, 2000 m.y.

IX The Origin and Nature of Early Life

Until the last decade, the records of Precambrian organic remains that were generally accepted as valid could almost have been counted on the fingers of one hand. Now, such records are too numerous to catalogue and, what is perhaps more important, they are spread through the whole span of geological history. This transformation has come about largely as a result of improved techniques for dealing with bodies of the order of a few micrometres in diameter.

The role of carbon is crucial to the evolution of organic compounds and hence of life itself. It seems probable that a *prebiological* phase in which simple organic compounds were built up without the agency of living organisms led to the earliest modifications of the atmosphere. Rutten (1962) has suggested that unchecked ultraviolet radiation promoted *inorganic photosynthesis* in the anoxygenic atmosphere, resulting in the dissociation of water and the consequent release of oxygen, hydrogen, nitrogen, sulphur and phosphorus. Berkner and Marshall (1965) concluded that the shielding effect of ozone would halt these processes at a very early stage but Brinkmann, among others, considers that such inorganic processes could have produced up to 25 per cent of the present oxygen content of the atmosphere. Nevertheless the bulk of the oxygen in our atmosphere must be biogenic, that is, the result of *organic photosynthesis*, and the newly discovered fossil record suggests that this process had begun over 3000 m.y. ago.

All forms of life appear to utilise many of the same chemical compounds in the same processes, governed by the same chemical coding. Inorganic reactions between water and simple compounds of carbon, hydrogen and nitrogen can be achieved experimentally with the production of amino acids, fatty acids, hydrocarbons and nitrogen compounds. The *carbonaceous meteorites* – a small group of meteorites characterised by the occurrence of hydrous silicates, and by a high content of water and organic matter – contain a variety of complex organic molecules as well as 'organised elements' up to about 30μm in diameter whose status is uncertain. If such compounds were formed inorganically in the early atmosphere, some means of concentration would seem to be required before the establishment of life became possible. Coacervation is one of the concentration mechanisms which have been considered in this context. Oparin, for example,. has experimented with proteins which under certain conditions, assemble into droplets with definite boundaries. Droplets and 'microspheres' formed in some such way exhibit some of the attributes of living organisms – for instance, they may selectively absorb substances and proliferate spontaneously. They bear a strong resemblance to the minute 'organised bodies' of certain carbonaceous meteorites and of certain early sediments such as the Bulawayan.

The geological environment in which such bodies passed into a truly living condition were, presumably, those in which the products of prebiological synthesis were concentrated in favourable conditions. Although geologists have been inclined to look to the sea, it is perhaps more probable that life originated in small bodies of water, especially those heated nutrient-rich waters held in rock-crevices or at the surface around volcanic centres. It is perhaps more than an accident of preservation that many of the earliest recorded organisms come from cherts associated with volcanic sequences.

The earliest organised bodies which are generally regarded as remains of living organisms (Table 10.4) belong exclusively to the category of *Procaryotes* which is represented at the present day by the bacteria and the blue green algae. These

Table 10.4. RECORDS OF PRECAMBRIAN ORGANISMS
(modified after P. Echlin, 1970)

Formation, approx. age	PROCARYOTES					EUCARYOTES		Organic compounds
	Bacteria	Unicells	Colonial	Filamentous	Complex filamentous	Algae	Fungi	
Bitter Springs cherts central Australia 1200-1000 m.y.	X	X	X	X	X	X	X	
Belt Series, western U.S.A.	X?	X	X				X?	
Huronian, Minnesota, U.S.A.		X	X	X				
Gunflint chert, Lake Superior Canada 2200-2000 m.y.	X	X	X	X	X		X?	X
Ketilidian, West Greenland			X?			X?		X
Witwatersrand Series South Africa c. 2600 m.y.	X?	X?				X		
Bulawayan Series, Rhodesia c. 2900								
Fig Tree Group South Africa 3200 m.y.	X	X						X

STROMATOLITES

organisms lack a distinct membrane-bound cell nucleus and do not possess a mechanism for cell division which allows for the accurate partition of and recombination of material from the nucleus. The elaborate provisions made by higher organisms for the handing-on and modification of genetic information are not fully developed in the Procaryotes, which appear to have evolved very little over the span of 3000 m.y. The earliest known *stromatolites*, built up by the trapping of lime by intertidal algal mats. are those of the Bulawayan, dated at nearly 3000 m.y. (p. 110): they differ little in essential characteristics from the corresponding structures of modern times and, like modern blue-green algae. probably practised photosynthesis involving the emission of oxygen. The tripartite micro-organism *Kakabekia*, first recognised in the Gunflint chert dated at over 2000 m.y., has turned up alive in ammoniacal soils below the walls of Harlech Castle in Wales.

The greatest single advance in organic evolution appears to have been the development of cells with a complex internal organisation in which the nucleus, bounded by a definite membrane, took on the role of a centre for the storage and transmission of genetic information. From this change arose the possibilities of sexual reproduction as a means of recombining genetic material, with a consequent enormous increase in the flexibility and adaptability of the processes of evolution. The *Eucaryotes*, in which the nucleus has this advanced role, are represented among fossil organisms in the Bitter Springs cherts of central Australia which are dated at about 1000 m.y. and possibly, though not certainly, in the Gunflint chert and other deposits dated at about 2000 m.y. (Table 10.4). All modern organisms with the exception of the bacteria, the blue-green algae and their allies, are Eucaryota and it seems certain that the attainment of this evolutionary grade in mid-Proterozoic times provided the stimulus for the subsequent diversification of the animal and plant kingdoms.

With the exception of the micro-organisms only accessible to study with high powered optical or electron microscopes, the dominant fossils of the Precambrian record are undoubtedly the *stromatolites*, the domed and lobate calcareous bodies built up in intertidal zones above mats of filamentous blue-green algae (Hoffman 1969). A broad variation in gross morphology has allowed stromatolites to be used as crude stratigraphical indices, especially by Russian and Canadian geologists. Keller and others, by means of serial sectioning of material from Asian successions, have established three assemblages of stromatolite forms which are considered to characterise three major Precambrian divisions dated at 1500 m.y., 1300–1000 m.y. and 900–850 m.y. Broad morphological transitions connect the assemblages and the whole series appears to express a long-drawn-out morphological evolution similar to that undergone by Phanerozoic organisms. The characters important in differentiating successive assemblages include the convexity of the laminae, the extent of branching and the details of lateral boundaries of the columns. Variations are not apparently related to conditions of sedimentation and are so constant that they have been used to zone Proterozoic strata in U.S.S.R. Comparable variations in the Denault Formation of Labrador have enabled Donaldson (1963) to recognise a succession of five zones.

X Unidirectional Trends in Geological History

On various grounds already discussed, it has seemed reasonable to accept a gradual transformation of the early *atmosphere and hydrosphere* leading through geological time to the compositions and volumes of the present-day atmosphere and hydrosphere. The processes characteristic of the surface of the litho-sphere – weathering, sedimentation and diagenesis – must, if this inference is correct, have operated in changing chemical environments, anoxygenic environments being characteristic at least of part of Archaean time and oxygenic atmospheres at least of the last 1500 m.y. Especially important here are certain sediments such as the iron formations which are sensitive to the state of oxidation. The changes with time in the lithology of ferruginous sediments have been linked by many geologists with the appearance and development of life.

Several proposals concerning the characters of early sediments have depended on the ratios of ferric to ferrous oxides. As we have already mentioned, the occurrence of pyrite and uraninite grains considered to be detrital has been cited by some as evidence that early sediments such as the Witwatersrand bankets and the Huronian conglomerates of Blind River in Ontario accumulated under reducing conditions. On the other side of the balance, the occurrence of detrital sediments carrying a hematitic cement is plausibly taken to indicate diagenesis under oxidising conditions. *Red beds* of this type appear in the stratigraphical record about mid-way through the Proterozoic with the Jotnian Sandstone of the Baltic shield, *c.*1400 m.y. (p. 38), the Vindhyan of India *c.*1400 m.y. (p. 134) and a considerable number of somewhat younger formations.

The *banded iron formations* associated with cherts occupy a crucial position in the array of iron-bearing sediments, since they are widely distributed in space but rather closely restricted in time. As we have seen, banded iron formations of major importance are recorded from every continent. Some (the Algoma type) are incorporated in Archaean provinces where they are associated with volcanics. Most are components of early Proterozoic successions deposited unconformable on, and especially around, the margins of stabilised Archaean cratons. Most of these early Proterozoic formations fall in the time-range 2300–2000 m.y. Virtually none is younger than 1800 m.y. Sedimentary iron formations of later Proterozoic and Phanerozoic successions differ fundamentally from the banded iron formations in occurrence, mineral constitution and chemistry, as is illustrated by Fig. 10.4 based on Govett.

The remarkable characters of the banded iron formations have been mentioned repeatedly, especially in connection with our account of the Hamersley Group in Western Australia (pp. 145–8). Further discussion of the environment in which these formations accumulated will be found in papers by James (1972), Govett (1966) and Trendall (1968). All authors are agreed that the deposits are chemical sediments and strong arguments have been advanced to suggest that they accumulated in partially enclosed or even lacustrine basins. The common occurrence of hematite indicates that oxygen was available during deposition, or early diagenesis, and provides the most important aspect of the formations in the present context. Preston Cloud has phrased the geochemical problem connected with the formations as 'how to transport the iron in solution under oxidising conditions, or to precipitate it under anoxidising conditions.' His

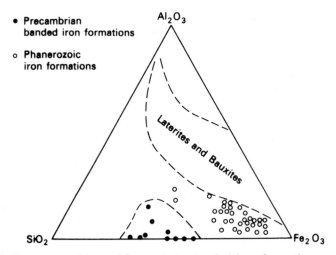

Fig. 10.4. The compositions of Precambrian banded iron formations compared with those of other types of sedimentary iron ore (after Govett, 1966)

solution of this problem involved the idea of a balanced relationship between the chemical precipitation of iron and the oxygenic activities of primitive organisms (1971). Iron was transported to the basin in the ferrous state and oxidised at the site of deposition by oxygen released during photosynthesis. The breakdown of this balance and the consequent termination of deposition of banded iron formations was the result of evolutionary advances enabling the organisms concerned to tolerate and subsequently make use of free oxygen. These concepts are expressed diagrammatically in Fig. 10.5.

The idea of a chemical evolution of the lithosphere was implicit in our discussion of the evolution of the earth's atmosphere and hydrosphere, since the prime role in the transfer of volatiles from the interior was assigned to the processes of magmatism. In the light of modern ideas on the importance of partial melting in the generation of magmas, it must be inferred that the mantle and perhaps the lower parts of the continental crust have been progressively depleted not only in volatiles but also in other fusible components.

Attempts to explore these ideas by geochemical investigations of provinces of different ages or by comparisons of specific rock types formed at different periods have, however, yielded results which are ambiguous. We have referred in an earlier chapter to suggestions that the earliest basic volcanics of greenstone belts older than 3000 m.y. were more magnesian than corresponding rocks of younger sequences (p. 101). Early granites have been discussed by many authors, notably by Eskola who believed that granites were generated in the Karelide and earlier orogenies on a scale never attained in later cycles. Mehnert has suggested that the older parts of the crust show unique developments of alkalies, especially of sodium, and concludes that the present distribution of these mobile elements is a consequence of one-way migration from the primary source. This concept is borne out to some extent by Engel's plots of potassium—sodium ratios in rocks of differing ages, where a continuous build-up in the relative

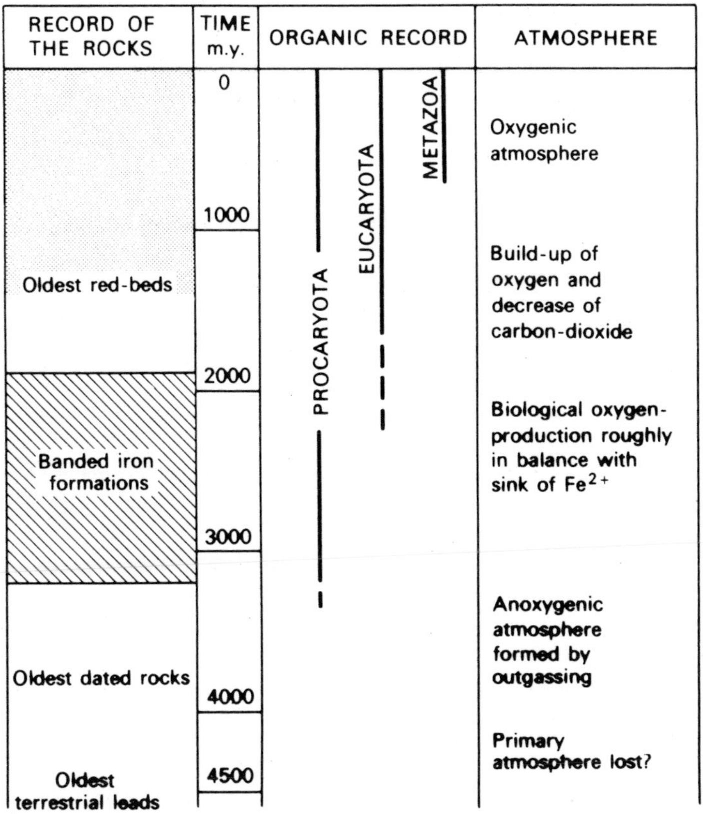

RECORD OF THE ROCKS	TIME m.y.	ORGANIC RECORD	ATMOSPHERE
	0	METAZOA / EUCARYOTA / PROCARYOTA	Oxygenic atmosphere
Oldest red-beds	1000		Build-up of oxygen and decrease of carbon-dioxide
	2000		Biological oxygen-production roughly in balance with sink of Fe^{2+}
Banded iron formations	3000		
Oldest dated rocks	4000		Anoxygenic atmosphere formed by outgassing
Oldest terrestrial leads	4500		Primary atmosphere lost?

Fig. 10.5. Changes in the earth's atmosphere in relation to the evolution of life (after Cloud, 1971)

importance of potassium appears to be abruptly reversed in Mesozoic times. Finally, we have noted that anorthosites have been assigned a special role in the early development of the crust by some authors.

When all these suggestions have been mentioned, however, one is left with a feeling of insufficiency. We still know little about the effects of variations of pressure and temperature in the source-regions on the composition of the magmas generated, still less about the geochemical effects of regeneration of crustal rocks and nothing about the composition or volume of the Precambrian oceanic crustal plates. While these unknowns remain, it seems fruitless to speculate further on the chemical evolution of the lithosphere as a whole.

A more limited question of interest in the present context is that of possible variations of the type of mineralisation with time, especially in so far as it concerns ore-deposits associated with the magmatism of the geological cycles. From the surveys of the record down to about 800 m.y given in the foregoing chapters, marked differences between the mineral deposits of different cycles have emerged. Some interim comments may be made on the contrasted styles of mineralisation in Archaean and Proterozoic cycles. Holmes drew attention to this contrast in his classic paper of 1951, especially as displayed in Africa, pointing out that gold ores were almost confined to the oldest cycles. The impressive list of Archaean provinces exhibiting gold mineralisation is given in Table 10.5 No subsequent period of 800 m.y. duration has seen the emplacement of gold on anything approaching the same scale.

Table 10.5. GOLD MINERALISATION IN ARCHAEAN PROVINCES

North America	southern part of Superior province Slave province, in or near low grade greenstone belts (>2500 m.y.)
Siberia	Aldan massif, stabilised about 2000 m.y.
Southern Africa	Transvaal massif, in low-grade greenstone belts of Swaziland Series (>3200 m.y.) Witwatersrand Series, placer-deposits of the Rand (2800–2300 m.y.) Rhodesia, in and near low-grade schist belts.
East Africa	Tanzania massif, in and near low-grade Nyanzian-Kavirondian belts.
West Africa	in Ghana, associated with granites in Birrimian volcanics and metasediments (c. 2000 m.y.)
India	Dharwar province, mainly in basic volcanics of medium or low grade (c. 2400 m.y.)
Australia	Yilgarn block, in and near low-grade volcanics of greenstone belts (>2400 m.y.)
South America	Guyana shield, near contacts of granites with low-grade basic volcanics of Barama-Mazaruni assemblage (>2000 m.y.)

Mineralisation in the chelogenic cycle 2200–1100 m.y. was not dominated by any single metal. Different continents display a variety of mixed ores, mainly sulphides of copper, zinc, arsenic, bismuth, molybdenum, tungsten, tin and uranium. Gold is nowhere of much importance, even in areas such as the western part of the Churchill province which incorporates large masses of regenerated Archaean greenstones. We must infer that some metals which were concentrated during one cycle were dispersed during a later cycle.

Finally, we may mention briefly some speculations concerning long term changes in the physical environments of the continental crust. We have already dealt with one aspect of this subject, the changing relationships between stable (cratonic) and mobile crustal regimes and have seen that the first true cratons appear to have been defined towards the end of the Archaean in the time-period 2800–2400 m.y. (pp. 171–5). The occurrence of a few kimberlites and alkaline centres dating back to mid-Proterozoic times (pp. 109, 119) suggests that environments in the magmatic source-regions of the mantle beneath the early cratons had much in common with those in force beneath the Phanerozoic cratons (the occurrence of detrital diamonds in formations dated at over 2000 m.y. in the Ivory Coast, and in the Roraima formation of Guyana (>1800 m.y.) suggests the interesting possibility that kimberlite emplacement began in Archaean times).

Some authors have concluded that the characters of Archaean complexes indicate the dominance of steep geothermal gradients in the Archaean crust.

There are, indeed, a number of records (for example in Rhodesia) of regionally developed assemblages characterised by andalusite in pelitic rocks, which suggest that the low-pressure metamorphic facies series is widely represented. The rapid transitions from medium or high, to low metamorphic grades in the lowest parts of the supracrustal successions in many greenstone belts suggest steep thermal gradients. But the fact that very low-grade supracrustals occur in deep infolds between gregarious batholiths suggests that the pattern of isograds is related to the rising granitic domes rather than to regional thermal 'highs'. Indications of regional metamorphism at high temperatures are provided by extensive terrains of granulite facies such as those in the pre-Ketilidian massif of Greenland. Parts of this massif, and of the corresponding Scourian complex of north-west Scotland, carry kyanite in pelitic gneisses, suggesting metamorphism related to the intermediate facies series.

The early Proterozoic Hudsonian belts of North America and the Svecofennides of Europe are characterised over very wide areas by a low-pressure facies series in which andalusite is sometimes accompanied by cordierite. The extent of assemblages of this type seems to imply the existence of steep geothermal gradients in the mobile belts during early Proterozoic times. Mobile belts formed over the same time-span in other regions, however, are characterised by the intermediate facies series. It is doubtful whether any general progression from steeper to less steep geothermal gradients can be made out, for the low-pressure facies series is developed on a considerable scale in several Phanerozoic mobile belts. There are, so far as we are aware, no records of the occurrence in rocks older than 1000 m.y. of the glaucophane bearing (blueschist) assemblages characteristic of the high-pressure facies series. This series is of considerable importance in Mesozoic and Tertiary mobile belts but only of local occurrence in Palaeozoic and late Precambrian belts. It remains to be established whether the apparent increase in the extent of the high-pressure trend of metamorphism was real or is an effect of preservation.

Bibliography Part I

The reference lists that follow are intended to supplement the information given in successive chapters as well as to cover some points of detail referred to in the text and figures. They are not comprehensive and certainly do not cover all the sources consulted during the writing of this book; but the information given is intended to provide ways into the literature by the citation of a number of reviews and other general works which have comprehensive bibliographies. Where papers are referred to more than once, the full reference is given in the first chapter list; later lists refer back to this chapter.

The maps listed below are general compilations showing a variety of geological features. Their cost will put them beyond the means of some readers but many of them can be consulted at major geological libraries.

Maps

EURASIA
Tectonic map of Europe (16 sheets) 1:2500 000
  Commission for the tectonic Map of the World, Moscow, 1964.
International Geological Map of Europe (2 sheets) 1:5000 000
  Commission for the geological Map of the World, UNESCO, 1971.
Metallogenic map of Europe (9 sheets) 1:2500 000
  Commission for the geological Map of the World, UNESCO (first sheet issued 1968)
Tectonic map of Eurasia (12 sheets) 1:5000 000
  Geol. Inst. Akad. Nauk SSSR, 1966.
Geological map of Britain (2 sheets) 1:625 000
  Institute of Geological Sciences, London.
Tectonic map of Britain 1:1000 000
  Institute of Geological Sciences, London 1966.

GREENLAND
Tectonic map of Greenland 1:2500 000
  G.G.U. Copenhagen.

NORTH AMERICA
Tectonic map of North America (2 sheets) 1:5000 000
  United States Geological Survey, 1969.
Basement map of North America 1:5000 000
  U.S.G.S. and American Association of Petroleum Geologists, 1967

Geological map of Canada 1:5000 000
  Geological Survey of Canada, Map 1250A, 1968.
Isotopic age map of Canada 1:5000 000
  Geological Survey of Canada, Map 1256A, 1969.
Aeromagnetic map of Canadian shield 1:5000 000
  Geological Survey of Canada.
Glacial map of North America 1:4555 000
  Geological Society of America, 1945.

SOUTH AMERICA
Carte géologique de l'Amerique du Sud (2 sheets) 1:5000 000
  Geological Society of America, 2nd Ed. 1964.

AFRICA
International tectonic map of Africa (9 sheets) 1:5000 000
  Association des Services géologiques d'Afrique and UNESCO 1968.
Gravity map of South Africa (4 sheets) 1:1000 000
  Bureau of Mineral Resources, South Africa, 1958.

AUSTRALIA
Geological map of Australia and Oceania 1:5000 000
  Commission for the geological Map of the World.
Tectonic map of Australia 1:2534 400
  Bureau of Mineral Resources, Canberra, 1960.
Metallogenic map of Australia and Papua New Guinea 1:5000 000
  Bur. Min. Res. Canberra (1972), Bull. 145.

1 The Geological Record

Ager, D. V. (1963). *Principles of Palaeoecology*, McGraw-Hill.
Aubouin, J. (1965). *Geosynclines,* Elsevier.
Bullard, E. C. (1968). The Bakerian Lecture: Reversals of the Earth's magnetic field. *Phil. Trans. Roy. Soc. Lond.*, A263, 481.
Dearnley, R. (1966). Orogenic fold-belts and a hypothesis of earth evolution. *Physics and Chemistry of the Earth*, 7, Pergamon.
Donovan, D. T. (1966). *Stratigraphy: An Introduction to Principles*, Murby.
Gastil, G. (1960). Continents and mobile-belts in the light of mineral dating. *Int. Geol. Cong. 21st Session*, Part 9, 162.
Gass I., *et al.* (eds) (1971). *Understanding the Earth,* Artemis.
Gegnoux, M. (1955). *Stratigraphic Geology* (trans. G. G. Woodford), Freeman.
Harland, W. B., *et al.* (eds) (1964). The Phanerozoic time-scale. *Q. Jl. geol. Soc , Lond.*, 120S.
Holmes, A. (1965). *Principles of Physical Geology* (2nd edition), Nelson.
Kummel, B. (1970). *History of the Earth* (2nd edition), Freeman.
Middlemiss, F., *et al.* (eds) (1971). Faunal provinces in space and time. *Geol. J.*, Special Issue 4.

Moorbath, S. (1967). Recent advances in the application . . . of radiometric age-data. *Earth Sci. Reviews*, **3**, 111.

Quennell, A. M. and E. G. Haldemann (1960). On the subdivisions of the Precambrian. *Int. geol. Congr. Proc.*, **9**, 170.

Runcorn, S. K. (1962). Convection currents in the earth's mantle. *Nature, Lond.*, **195**, 311.

Sutton J. (1963). Long-term cycles in the evolution of the continents. *Nature, Lond.*, **198**, 731.

—— (1969) 'Rates of change within orogenic belts' *Time and Place in Orogeny*, Geol. Soc. Lond. special publications 3, 239.

Code of Stratigraphic Nomenclature, 1961, *Bull. Am. Ass. Petrol. Geol.*, 45.

Report of Stratigraphical Code Sub-Committee, 1967, *Proc. geol. Soc. Lond.*, No. 1638.

2 European Shield-areas

Backlund, H. G. (1936). Der 'Magmaaufsteig' in Faltengebirge. *B.G.F.*, **115**, 293.

Bogdanoff, A. A., M. V. Mouratov and N. S. Schatsky (eds.) (1964). *Tectonique de l'Europe* (explanation of Tectonic Map of Europe) Moscow.

Edelman, N. (1960). The Gulkrona region, south-west Finland. *Bull. Comm. géol. Finlande*, 137.

Eskola, P. (1948). The problem of mantled gneiss domes. *Q. Jl. geol. Soc. Lond.*, **104**, 461.

—— (1952). On the granulites of Lapland. *Am. J. Sci.*, (Bowen Vol.), 133.

—— (1963). The Precambrian of Finland, In *The Precambrian*, Vol. 1, (ed. Rankama), Interscience.

Heier, K. S. and W. Compston (1969). Interpretation of Rb—Sr age patterns in high-grade metamorphic rocks, north Norway. *Norsk geol. Tidssk.*, **49**, 257.

Holtedahl, O. (ed.) (1960). *Geology of Norway. Norges Geol. Unders.* No. 208.

Magnusson, N. H. (1965). Precambrian history of Sweden. *Q. Jl. geol. Soc. Lond.*, **121**, 1.

Nalivkin, D, V. (trs. N. Rast) 1973. *Geology of the U.S.S.R.*, Oliver and Boyd.

Polkanov, A. A. and E. K. Gerling (1960). The Precambrian geochronology of the Baltic shield. *Int. Geol. Congr. 21st Session*, 9, 183.

Read, H. H. (1957). *The Granite Controversy*, Murby.

Sederholm, J. J. (1930). The Pre-Quaternary rocks of Finland. *Bull. Comm. géol. Finlande*, 91.

Semenenko, N. P. (1968). Geochronology of the Ukrainian Precambrian. *Can. J. Earth Sci.*, **5**, 661.

Simonen, A. (1953). Stratigraphy and sedimentation of the Svecofennidic, early Archaean supracrustal rocks in south-western Finland. *Bull. Comm. géol. Finlande*, 160.

—— (1960). Pre-Quaternary rocks in Finland. *Bull Comm. géol. Finlande*, 191.

Smithson, S. (1965). The nature of the 'granitic' layer of the crust in the southern Norwegian Precambrian. *Norsk geol Tidssk.* **45**, 113.

Starmer, I. C. (1972). The Sveconorwegian regeneration and earlier orogenic events in the Bamble series, South Norway. *Norges Geol. Unders*, **273**, 37.

Tuttle, O. F. and N. L. Bowen (1958). Origin of granite in the light of experimental studies in the system $NaAlSi_3O_8-KAlSi_3O_8--SiO_2-H_2O$. *Geol. Soc. Am. Mem.*, 74.

Wegmann, C. E. and E. H. Kranck (1931). Beiträge zur Kenntnis der Svecofenniden in Finnland. *Bull. Comm. géol Finlande*, 189.

Wetherill, G. W., O. Kouvo, G. R. Tilton and P. W. Gast (1962). Age measurements on rocks from the Finnish Precambrian. *J. Geol.*, **70**, 74.

3 North Atlantic Shield-areas

Black, L. P., N. H. Gale, S. Moorbath, R. J. Pankhurst and V. R. McGregor (1971). Isotopic dating of very early Precambrian amphibolite facies gneisses from the Godthåb district, West Greenland. *Earth and Plan. Sci. Letters*, **12**, 246.

Bowes, D. R., B. C. Baraooah and S. G. Khoury (1971). Original nature of Archaean rocks of north-west Scotland. *Spec. Pap. geol. Soc. Aust.*, No. 3, 77.

Bridgwater, D., A. Escher and J. Watterson (1973). Tectonic displacements and thermal activity in two contrasting Proterozoic mobile belts from Greenland. *Phil. Trans. R. Soc.*, A273, 513.

———, J. Watson and B. F. Windley (1973). The Archaean craton of the North Atlantic region. *Phil. Trans. R. Soc.*, A273, 493.

——— and others (1974). A horizontal tectonic regime in the Archaen of Greenland, etc. *Precambre. Res.*, I.

Coward, M. P., P. W. Francis R. H. Graham and J. Watson (1971). Large-scale Laxfordian structures of the Outer Hebrides in relation to those of the Scottish mainland. *Tectonophysics*, **10**, 425.

Craig, G. Y. (ed.) (1965). *The Geology of Scotland*, Oliver and Boyd.

Dearnley, R. (1962). An outline of the Lewisian complex of the Outer Hebrides in relation to that of the Scottish mainland. *Q. Jl. geol. Soc. Lond.*, **118**, 143.

Higgins, A. K. (1970). The stratigraphy and structure of the Ketilidian rocks of Midaternes. south-west Greenland. *Med. om Grøn.*, **189**, No. 2.

McGregor V. R. (1973). The early Precambrian gneisses of the Godthåb region. *Phil Trans. Roy. Soc. Lond.*, A273, 343.

Moorbath, S., H. Welke and N. H. Gale (1969). The significance of lead isotope studies in ancient, high-grade metamorphic basement complexes, etc. *Earth and Plan. Sci. Letters*, **6**, 245.

Peach, B. N., J. Horne *et al.* (1907). The geological structure of the North West Highlands of Scotland. *Mem. geol. Surv. Scotl.*

Sheraton, J. W. (1970). The origin of the Lewisian gneisses of north-west Scotland with particular reference to the Drumbeg area, Sutherland. *Earth and Plan. Sci. Letters*, **8**, 301.

Sutton, J. and J. Watson (1951). The pre-Torridonian metamorphic history of the Loch Torridon and Scourie areas in the North-west Highlands, etc. *Q. Jl. geol. Soc. Lond.*, **106**, 241.

Watterson, J. (1968). Homogeneous deformation of the gneisses of Vesterland, South-west Greenland. *Med. om Grøn.*, **175**, No. 6.

Wegmann, C. E. (1935). Zur Deutung der Migmatite, *Geol. Rundsch*, **26**, 307.

Windley, B. F., R. K. Herd and A. A. Bowden (1972). The Fiskenaesset complex, West Greenland, Part 1. *Bull. Grøn. Geol. Unders.*

―――― and D. Bridgwater (1971). The evolution of Archaean low- and high-grade terrains. *Spec. Pap. geol. Soc. Aust.*, **3**, 33.

4 Precambrian of the North American Craton

Adams, F. D., and A. E. Barlow (1910). Geology of the Haliburton and Bancroft areas, Province of Ontario. *Geol. Surv. Can.* Mem. 6.

Baragar, W. R. A., (1967). Wakauch Lake map area, Quebec-Labrador. *Geol. Surv.* Can., Mem. 344.

Davidson, A. (1972). The Churchill province. Variations in tectonic styles in Canada. *Spec. Pap. geol. Assoc. Can.*, **11**, 381.

Dimroth, E., 1970. Evolution of the Labrador geosyncline. *Bull geol. Soc. Am.*, **81**, 2717.

Donaldson, J. A. *et al.* (1973) Drift of the Canadian shield. *Implications of continental drift to the Earth Sciences*, Vol. 1, 3.

Fahrig, W. F., 1961. The geology of the Athabasca formation. *Bull. geol. Surv. Can.*, 68.

Fenton, M. D., and G. Faure (1969). The age of the igneous rocks of the Stillwater complex of Montana. *Bull. geol. Soc. Am.* **80**, 1599.

Gastil, G. and others (1960). The labrador geosyncline. *Int. geol. Congr. Proc.*, **9**, 21.

Goodwin, A. M. (1968). Evolution of the Canadian shield. *Proc. geol. Ass. Can.*, **19**, 1.

―――― (1971). Metallogenic patterns and evolution of the Canadian shield. *Spec. Pap. geol. Soc. Aust.*, **3**, 157.

Green. D. C., H. Baadsgaard and G. L. Cumming (1968). Geochronology of the Yellowknife area, North-west Territories, Canada. *Can. J. Earth Sci.*, **5**, 725.

Hoffman. P. (1973). The Coronation geosyncline: Lower Proterozoic analogue of the Cordilleran geosyncline in the north-western Canadian shield. *Phil. Trans. R. Soc.*, A, **273**, 547.

King, E. R. and I. Zietz (1971). Aeromagnetic study of the mid-continent gravity high of central United States. *Bull. geol. Soc. Am.*, **82**, 2187.

Muhlberger, W. R., R. E. Denison and E. G. Lidiak (1967). Basement rocks in continental interior of United States. *Bull. Am. Ass. Petrol. Geol.*, **51**, 2351.

Payne, A. V. *et al.* (1964). A line of evidence supporting continental drift. *I.U.G.S. circ. letter*, **13**, 75.

Pettijohn, F. J. (1957). Palaeocurrents of Lake Superior Precambrian quartzites. *Bull. geol. Soc. Am.*, **68**, 469.

Quirke, T. T. and W. H. Collins (1930). The disappearance of the Huronian. *Geol. Surv. Can.*, Mem. 160.

Roscoe, S. M. (1968). Huronian rocks and uraniferous conglomerates of the Canadian shield. *Geol. Surv. Can.* Paper 68–40.

Stockwell, C. H. (1968). Geochronology of stratified rocks of the Canadian shield. *Can. J. Earth Sci.*, 5, 693.
—— (1972). Revised Precambrian time-scale for the Canadian shield. *Geol. Surv. Canada*, Paper 72–52.
Tilley, C. E. (1958). Problems of alkali rock genesis. *Q. Jl. geol. Soc. Lond.*, 113, 323.
Walton, M. and D. de Waard (1963). Orogenic evolution of the Precambrian in the Adirondack highlands, a new synthesis. *Proc Kon. Ned. Acad. Wetensch.*, 66, 98.
Watson, J., 1973. Effects of reworking on high-grade gneiss complexes. *Phil. Trans. Roy. Soc. Lond.*, A273, 443.

5 Precambrian of Asiatic Laurasia

Kazakov, G. E. and K. G. Knorre (1970). Geochronology of the Upper Precambrian of the Siberian platform, Uchur-Maja region. *Eclog. geol. Helv.*, 63, 173.
Keller, M. (1964). The Riphean group. *Int. geol. Congr. 22nd Session*, Part 10, 323.
Nalivkin, D. V., 1973 (Trs. N. Rast). *Geology of the U.S.S.R.* Oliver & Boyd.
Sokolov, B. S. (1964). The Vendian and the problem of the boundary between the Precambrian and the Palaeozoic group. *Int. geol Congr. 22nd Session*, Part 10, 288.
Tougarinov, A. I. (1968). Geochronology of the Aldan shield, south-eastern Siberia. *Can. J. Earth Sci.*, 5, 649.

6 The African Cratons

Allsopp, H. L., T. J. Ulrych and L. O. Nicolaysen (1968). Dating some significant events in the history of the Swaziland System by the Rb–Sr isochron method. *Can. J. Earth Sci.*, 5, 605.
Anhaeusser, C. R., R. Mason, M. J. Viljoen and R. P. Viljoen (1969). Reappraisal of some aspects of Precambrian shield geology. *Bull. geol Soc. Am.*, 80, 2175.
Cahen, L. and N. J. Snelling (1968). *The geochronology of equatorial Africa.* North-Holland.
—— and J. Lepersonne (1967). The Precambrian of the Congo, Ruanda and Burundi. *The Precambrian*, Vol. 3 (ed. Rankama), Interscience.
Clifford, T. N. (1966). Tectonometallogenic units and metallogenic provinces of Africa. *Earth and Plan. Sci. Letters* 1, 421.
—— (1970). The structural framework of Africa *in African magmatism and tectonics*, (ed. Rayner), Oliver and Boyd.
Combe, A. D. (1932). The geology of south-west Ankole. *Geol. Surv. Uganda*, Mem. 2.
Davidson, C. F. (1965). The mode of origin of banket orebodies. *Bull. Inst. Min. Metal.*, 74, 319.

Du Toit, A. L. (1939). *The Geology of South Africa*, 2nd edit., Oliver & Boyd.

Hall, A. L. (1932). The Bushveld igneous complex of the central Transvaal. *Union S. Africa geol. Surv.*, Mem. 28.

Hepworth, J. V. (1967). The photogeological recognition of ancient orogenic belts in Africa. *Q. Jl. geol. Soc. Lond.*, 123, 253.

Holmes, A. (1951). The sequence of Precambrian orogenic belts in South and Central Africa. *Int. Geol. Congr. 18th Session.* 14, 254.

Hunter D. R. (1970). The ancient gneiss complex in Swaziland. *Trans. geol. Soc. S. Africa*, 73, 107.

Kennedy, W. Q. (1965). The influence of basement structure on the evolution of the coastal (Mesozoic and Tertiary) basins of Africa; *In: Salt basins around Africa*, Inst. Petrol.

MacGregor, A. M. (1951). Some milestones in the Precambrian of Southern Rhodesia. *Trans. geol. Soc. S. Africa*, 54, xxvii.

Piper, J. D. A. (1973). Geological interpretation of palaeomagnetic results from the African Precambrian. *Implications of continental drift to the Earth Sciences*, Vol. 1, 19.

Pretorius, D. A. (1964). The geology of the central Rand goldfield. *In: The geology of some ore deposits of southern Africa*, Vol. 1, (ed. Haughton), *Geol. Soc. S. Africa.*

Shackleton, R. M. (1970). On the origin of some African granites. *Proc. geol. Ass.*, 81, 549.

Stowe, C. W. (1968). The geology of the country south and west of Selukwe. *Bull. geol. surv. Rhod.*, 59.

Talbot, C. J. (1968). See p. 201.

Vail, J. R. (1970). Tectonic control of dykes and related irruptive rocks in eastern Africa; *In: African magmatism and tectonics* (ed. Rayner), Oliver and Boyd.

Viljoen, R. P. and M. J. Viljoen (1971). The geological and geochemical evolution of the Onverwacht Volcanic Group of the Barberton Mountainland, South Africa. *Spec. Pap. geol. Soc. Aust.* 3, 133.

Worst, B. G. (1960). The Great Dyke of Southern Rhodesia. *Bull geol. Surv. S. Rhod.*, 47.

7 The Indian Craton

Cooray, P. G. (1962). Charnockites and their associated gneisses in the Precambrian of Ceylon. *Q. Jl. geol. Soc. Lond.* 118, 239.

Crawford, A. R. (1969). Reconnaissance Rb–Sr dating of the Precambrian rocks of southern Peninsular India. *J. geol. Soc. India*, 10, 117.

—— and R. L. Oliver (1968). The Precambrian geochronology of Ceylon. *Spec. Pap. geol. Soc. Aust.*, 2, 283.

Heron, A. M. (1953). The geology of central Rajputana. *Mem. geol. Surv. India*, 79.

Holland. T. H. (1900). The charnockite series. *Mem. geol. Surv. India*, 28.

Howie, R. A. (1955). The geochemistry of the charnockite series of Madras, India. *Trans. Roy. Soc. Edin.*, 62, 725.

Naha, K., A. K. Chaudhuri and P. Mukherji (1967). Evolution of the banded gneissic complex of central Rajasthan, India. *Contr. Mineral and Petrol.* 15, 191.

Pichamuthu, C. S. (1965). Regional metamorphism and charnickitisation in Mysore State, India. *Indian Min.*, 6, 119.

Sarkar, S. N., and A. K. Saha (1963). On the occurrence of two intersecting Precambrian orogenic belts in Singbhum and adjacent areas, India. *Geol. Mag.*, 100, 69.

———— , A. K. Saha and J. A. Miller (1969). The geochronology of the Precambrians of Singbhum and adjacent regions, eastern India. *Geol. Mag.* 106, 15.

Subramaniam, A. P. (1956). Mineralogy and petrology of the Sittampundi complex, Salem district, Madras State, India. *Bull. geol. Soc. Am.* 67, 317.

8 The Australian Craton

Arriens, P. A. (1971). The Archaean geochronology of Australia, *Spec. Pap. geol. Soc. Aust*, 3, 11.

Binns, R. A. (1964). Zones of progressive regional metamorphism in the Willyama complex, etc. *J. geol. Soc. Aust.*, 11, 283.

Carter, E. K., J. H. Brooks and K. R. Walker (1961). The Precambrian mineral belt of north-western Queensland. *Bull. Bur. Min. Res.*, 51.

Gellatly, D. C., G. M. Derrick and K. A. Plumb (1970). Proterozoic palaeocurrent directions in the Kimberley region, north-west Australia. *Geol. Mag.* 108, 249.

Glover, J. E. (ed.) (1971). Symposium on Archaean rocks held at Perth. *Spec. Pap. geol. Soc. Aust.*, 3.

Lambert, I. B. and K. S. Heier (1968). Geochemical investigations of deep-seated rocks in the Australian shield. *Lithos*, 1, 30.

McDougall, I., P. R. Dunn, W. Compston, A. W. Webb, J. R. Richards and V. M. Bofinger (1965). Isotopic age determinations in Precambrian rocks of the Carpentaria region, Northern Territory, Australia. *J. geol. Soc. Aust.*, 12, 67.

McLeod, W. N. (1966). The geology and iron deposits of the Hamersley Range area, Western Australia. *Bull. geol. Surv. W. Aust.*, 117.

Trendall, A. F. and J. G. Blockley (1970). The iron formations of the Precambrian Hamersley Group Western Australia. *Bull. geol. Surv. W. Aust.*, 119.

Williams, I. R. (1970). Explanatory notes on the 1:150 000 Kurnalpi Geological Map Sheet, *Geol. Surv., Western Australia Record*, 1970/71.

Wilson, A. F. (1969). Granulite terrains and their tectonic setting and relationship to associated metamorphic rocks in Australia. *Spec. Pap. geol. Soc Aust.*, 2, 243.

———— , W. Compston, P. M. Jeffery and G. H. Riley (1960). Radioactive ages from the Precambrian rocks in Australia. *J. geol. Soc. Aust.*, 6, 179.

9 The Cratons of South America and Antarctica

Adie, R. J. (1962). The geology of Antarctica. *Am. geophys. Union Geophys. Mon.*, No. 7, 26.

—— (ed.) (1972). Antarctic geology and geophysics. *I. U. G.S. Series B,* No. 1.

Choubert, B. (1956). French Guiana. *Mem. geol. Soc. Am.* **65.** 65.

Cordani, U. G., G. C. Melcher and F. M. K. de Almeida (1968). Outline of the Precambrian geology of South America. *Can. J. Earth Sci.,* **5,** 629.

Ebert, H. (1957). Beitrag zur Gliederung des Präkambriums in Minas Gerais. *Geol. Rundsch.,* **47,** 471.

Faure, G., J. G. Murtaugh and R. J. E. Montigny (1968). The geology and geochronology of the basement complex of the central Transantarctic Mountains. *Can. J. Earth Sci.,* **5,** 555.

Grabert, H. (1962). Zum Bau des Brasilianischen Schildes. *Geol. Rundsch.,* **52,** 292.

Harrington, H. J. (1962). Palaeogeographic development of South America. *Bull. Am. Ass. Petrol Geol.,* **46,** 1773.

Herz, N. (1970). Gneissic and igneous rocks of the Quadrilatero Ferrifero, Minas Gerais, Brazil. *Prof. Paper U.S. geol. Surv.,* 641-B.

Kalliokowski, J. (1965). Geology of north-central Guyana shield, Venezuela. *Bull geol. Soc. Am.,* **76,** 1027.

Kizaki, K. (1965). Geology and petrography of the Yamato Sanmyaku, East Antarctica. *Jap. Ant. Res. Exp. 1956–62, Sci. Rep. Series,* C, 3.

McConnell, R. B. and E. Williams (1970). Distribution and provisional correlation of the Precambrian of the Guiana shield. *Proc. 8th Guiana Conf.,* 1969.

Oliviera, A. I. de (1956). Brazil. *Mem. geol. Soc. Am.,* **65,** 3.

Segnit, E. R. (1957). Sapphirine-bearing rocks from MacRobertson Land Antarctica. *Miner Mag.,* **31,** 690.

Snelling, N. J. and R. B. McConnell (1969). The geochronology of Guyana. *Geol en Mijnb.,* **48,** 201.

Stillwell, F. L. (1918). The metamorphic rocks of Adelie Land. *Australasian Antarctic Exped. 1911–14, Sci. Rep. Series A,* 3.

10 Problems of the Precambrian Record

Berrange, J. P. (1965). Some critical differences between orogenic–plutonic and gravity-stratified anorthosites. *Geol. Rundsch.,* **55,** 617.

Berkner, L. V. and L. C. Marshall (1965). History of major atmospheric components. *Proc. Nat. Acad. Sci. U.S.A.,* **53,** 1215.

Briden, J. C. (1973). Applicability of plate tectonics to pre-Mesozoic time. *Nature, Lond.,* **244,** 400.

Bullard, E. C., J. E. Everett and A. G. Smith (1965). The fit of the continents around the Atlantic. *Phil. Trans. Roy. Soc. Lond.* Ser. A258, 41.

Cloud, P. (1971). The primitive earth *In: Understanding the earth* (ed. Gass *et al.*) Artemis Press.

Crook, K. A. W. (1966). Principles of Precambrian time-stratigraphy. *J. geol. Soc. Aust.,* **13,** 195.

Dearnley, R. (1965). Orogenic fold-belts, convection and expansion of the earth. *Nature, Lond.,* **206,** 1248.

—— (1966). see p. 193.

Donaldson, J. A. (1963). Stromatolites in the Denault Formation, Marion Lake, Coast of Labrador Newfoundland. *Bull. geol Surv. Can* 102.

Echlin, P. (1970). The origins of plants In: *Phytochemical phylogeny*, Academic Press.

Govett, G. J. S. (1966). Origin of banded iron formations. *Bull. geol. Soc. Am.*, 77, 1191.

Hoffmann, H. J. (1969). Attributes of stromatolites. *Geol. Surv. Canada*, Paper 69–39.

Holmes, A. (1951). see p. 198.

Rubey, W. W. (1951). Geologic history of sea water: an attempt to state the problem. *Bull. geol. Soc. Am.*, 62, 1111.

Rutten, M. (1962). *The geological aspects of the origin of life on earth.* Elsevier.

Saggerson, E. P. and L. M. Turner (1972). Some evidence for the evolution of regional metamorphism in Africa. *Int. geol. Congr. 24th Session*, 1, 153.

Shackleton, R. M. (1973). Correlation of structures across Precambrian orogenic belts etc. *Implications of Continental Drift to the Earth Sciences*, Academic Press, Vol. 2, 1091.

Stratigraphic classification and terminology, 2nd edn., Natl. Committee of Geologists of the U.S.S.R.

Sutton, J. (1963). Long-term cycles in the evolution of the continents. *Nature. Lond.*, 198, 731.

—— (1968). The extension of the geological record into the Precambrian. *Proc. geol. Assoc.*, 78, 493.

—— and J. Watson (1974). Tectonic evolution of continents in early Proterozoic times. *Nature, Lond.*, 247, 433.

Sylvester-Bradley, P. C. (1971). Environmental parameters for the origin of life. *Proc. geol. Assoc.*, 82, 87.

Talbot, C. J. (1968). Thermal convection in the Archaean crust? *Nature Lond.*, 220, 552.

Trendall, A. F. (1968). Three great basins of Precambrian banded iron formations: a systematic comparison. *Bull. geol. Soc. Am.*, 79, 1527.

—— (1972). Revolution in earth history. *J. geol. Soc. Aust.*, 19, 287.

Urey, H. C. (1952). *The planets: their origin and development.* Yale Univ. Press.

Watson, J. (1973). Ore-deposition in relation to crustal evolution. *Trans. Instn. Mining Metal*, 82, B107.

Index

Numbers in roman refer to pages in Part I; numbers in italic refer to Part II.